Back to the Future in the Caves of Kaua'i

Back to the Future in the Caves of Kaua'i

A Scientist's Adventures in the Dark

David A. Burney

Yale
UNIVERSITY PRESS
New Haven & London

Set in Monotype Bulmer by Duke & Company, Devon, Pennsylvania.
Printed in the United States of America by Sheridan Books, Ann Arbor, Michigan.

Library of Congress Cataloging-in-Publication Data
Burney, David A., 1950–
Back to the future in the caves of Kaua'i : a scientist's adventures in the dark / David A. Burney.
p. cm.
Includes bibliographical references and index.
ISBN 978-0-300-15094-0 (alk. paper)
1. Kauai (Hawaii)—Antiquities. 2. Excavations (Archaeology)—Hawaii—Kauai. 3. Caves—Hawaii—Kauai. 4. Paleoecology—Hawaii—Kauai. 5. Prehistoric peoples—Hawaii—Kauai. 6. Human ecology—Hawaii—Kauai—History. 7. Nature—Effect of human beings on—Hawaii—Kauai—History. 8. Time—Philosophy. 9. Kauai (Hawaii)—Environmental conditions. 10. Burney, David A., 1950– —Travel—Hawaii—Kauai. I. Title.
DU628.K3B87 2010
996.9'4102—dc22 2009049295

A catalogue record for this book is available from the British Library.

This paper meets the requirements of ANSI/NISO Z39.48-1992 (Permanence of Paper).

10 9 8 7 6 5 4 3 2 1

In memory of Eloise Nelson Pigott
(1925–2004)

What is it?

petrified Pleistocene sand pile
the plumbing of a sacred spring
subterranean tide-pool
fresh dark water in a coastal desert
big pickling jar full of old landscapes
millennium of garbage dumps
poachers' hideout
warriors' garrison
battlefield
place of divination
Chinese opium den
hippie commune
movie set
witches' gathering place
victim of tempest and tsunami
place of restoration
Poor Man's Time Machine

Makauwahi Cave.

Contents

Preface xi
Acknowledgments xiii

ONE
Time's Most Important Moment 1

TWO
Proverbial Tracks 10

THREE
Constructing a "Poor Man's Time Machine" 18

FOUR
Owl Omens 29

FIVE
Opening Ancient Doors 36

SIX
Characters and a Stage, but No Script 47

SEVEN
Fishponds 57

EIGHT
A Snails' Tale 64

NINE
Mauka Marshes 68

TEN
So What Happened, Anyway? 75

ELEVEN
Greetings from Old Kaua'i 86

TWELVE
Irrigating the Future 104

THIRTEEN
The Tour 118

FOURTEEN
Right Here, Right Now 130

FIFTEEN
Finding a Future in the Past 153

Glossary 173
Notes 175
Bibliography 181
Index 189

Preface

This is a book about time, although I don't claim to understand what time is. The physicists may be getting a handle on that, but it's not something I understand well enough to judge. I do claim, however, to know a little something about what time *does,* because I am a paleoecologist, a type of ecologist who studies past environments and the changes that climate and humanity have put them through. From three decades of thinking about this issue and studying ancient sites around the world, I am convinced that humans represent a kind of watershed in time on this planet. Something about human intelligence and how we use it to manipulate our environment is fundamentally different from anything that came before us. But why are humans so universally devastating for nature, particularly in the early stages of contact?

I want to tell three stories about time and humans as deliberate or unwitting environmental engineers. These three examples are all on quite different time scales. On a scale of tens of thousands of years, there is this big overriding question: *What is it about human arrival in any place that is so inevitably troublesome for nature?* The second story is about Kaua'i, where my wife, Lida, and I currently live, that northernmost of the inhabited islands of the Hawaiian Archipelago. *What has happened there in recent millennia and might happen in the near future, and what can this tell us about the rest of the planet?*

The third question is the one that will take me a whole book to answer, or maybe even more. *How is it that one place on Kaua'i, a particular cave, changed two people's lives so profoundly?*

This is an entirely true story, drawn from my field notebooks, articles I have published, and the memories of people whose lives have been touched by this special place, a place I often refer to as my "Poor Man's Time Machine."

Makauwahi Cave may be as close as some of us will ever get to time travel, I suspect. With hard work, one can coax out of this place a lot of information about the past. We know that for sure now, and this longer view can lead to a lot of thinking about the future. For nearly two decades, I have experienced in my research there a sense of the immensity and power of time that I had only known abstractly before. After sharing the place with thousands of visitors, I am convinced that it does the same thing to a lot of other people. Makauwahi Cave is only one place, but it has lived in many times. More than anything else, this book is about that place and those times.

Acknowledgments

I am in various ways indebted to all the thousands of people who have visited Makauwahi Cave in the course of the many years that Lida and I have been working there. This book is in a sense about them and the thousands more who I hope will pass that way in the future.

The manuscript was produced with support from the John Simon Guggenheim Memorial Foundation. I owe special thanks to the foundation and to my employer, the National Tropical Botanical Garden, for allowing me the time to do this. Chipper Wichman, the executive director and CEO of the botanical garden, deserves special thanks for being so supportive of this diversion from my regular duties.

To my wife, Lida Pigott Burney, I owe the most thanks, as she put in way more than half of the personal energy required to see Makauwahi Cave Reserve become a reality. To our grown-up children, Florence Lillian Mara Burney and James Alexander Pigott Burney, *mahalo* for being so patient with your parents' unusual lives, and for absorbing so well these unique experiences offered by our lives together as you grew up in exotic places. Thanks also for translating that into your own rapidly expanding lives in ways that make your parents proud.

All of the many scientists and conservationists who have worked at Makauwahi and our other sites deserve a lot of the credit for what we think we know about the cave, Kaua'i, and the extinction process. Drs. Storrs Olson and Helen James of the National Museum of Natural History (Smithsonian Institution) Bird Division and their children, Travis and Sydney, deserve a lot of thanks for their almost daily contributions as this project took shape from 1996 to 1999. The late William Kenji (Pila) Kikuchi, and his wife, Dolly, and their three daughters, Kristina, Kathleen, and Michelei, as well as Pila's cousin Katsuo Kikuchi, took us in like family and probably did more than anyone to convince us we should be living on Kaua'i. Our friends Reg and Sandy Gage had a hand in that as well, and Reg led us to many other good sites on Kaua'i and taught us about the local land-snail fauna.

Other scientists who moved us along in the process of discovery at Makauwahi include Fred Grady of Smithsonian Paleobiology, Dr. Julian Hume of the Natural History Museum (London), and more recently Dr. Nicholas Porch of the Australian National University. Thanks for being so useful in identifying difficult taxa and visualizing elements of past landscapes. Drs. Dan Livingstone, Ross MacPhee, and Paul Martin helped me as a graduate student to see the bigger picture concerning extinction and environmental change.

Thanks also to key scientific collaborators at the University of Hawaii. In addition to Pila, these have included Brian Yamamoto, Mike Kido, Ken Kaneshiro, and Terry Hunt. At Bishop Museum in Honolulu, we thank Frank Howarth for his guidance on how to manage blind cave invertebrates. At the National Tropical Botanical Garden, Ken Wood, Steve Perlman, Mike De Motta, Dave Lorence, Warren Wagner, and Bob Nishek have been great sources for information on rare Hawaiian plants.

This story would have ended abruptly well before the best parts if it had not been for the forbearance of the landowner, Grove Farm Company. Thanks to the support of Allan Smith, David Pratt, Mark Hubbard, Warren Haruki, Marissa Sandblom, and Mike Tresler of that firm, our continual presence and unusual activities at Makauwahi over the years have been not only tolerated but positively supported, even financially, by the company.

Five persons who played a large role down at the cave over the years are no longer with us: Adena Gillin, Dr. William Klein, Dr. Pila Kikuchi, LaFrance Kapaka-Arboleda, and Dave Boynton each gave us critical help along the way and will always be an important part of the history of this place. Makauwahi Cave Reserve is dedicated to their memory. We also thank the members of local organizations that have played a role in the cave's story, particularly Malama Maha'ulepu and the National Tropical Botanical Garden's volunteer organization, Na Lima Kokua. Special thanks go to our most regular volunteers, Mel Gable, Ed Sills, and Mary and Barry Werthwine. To Jimmy Miranda and the cowboys of CJM Stables and farmer Adam Killermann, thanks for being good neighbors to the Cave Reserve.

For the production of the book I owe special thanks to Lida, Mara, and Alec for reading it over and telling me the truth until I got it right. Down East in Carteret County, North Carolina, where I hid out while writing, close friends generally did me the favor of coming around and keeping my spirits up without coming around too often for me to get the work done. In this regard I especially thank Robin and Captain Dennis Chadwick, Brian Blake and Barbara Garrity-Blake, Allyn and JoAnn Powell, and Mike and Elizabeth Peeler. Prolific author and local storyteller Sonny Williamson read an early draft and warned me against "using too many big words." In hindsight, I think that was some of the best advice I received, and his wife Jenny's stewed hard crabs were the best, too.

The research projects discussed here have been supported by major grants from the National Science Foundation, the National Geographic Society, the National Oceanic and Atmospheric Administration (Human Dimensions of Global Change Program), and the Smithsonian Institution. Additional research support has come from Kaua'i Community College, the University of Hawaii, the Bette Midler Family Trust, the Kilauea Point Natural History Association, the Waipa Farmers' Cooperative, the National Park Service, and faculty research grants and fellowships from Fordham University. Our conservation projects have been supported by grants and contracts from the Wildlife Habitat Incentives Program of the Natural Resources Conservation Service (U.S. Department of Agriculture), the U.S. Fish and Wildlife Service, Grove Farm Company, and private donors and foundations.

Lida and I have been really lucky to live on Kaua'i. We appreciate the help and tolerance of its good people, and we especially owe a debt of gratitude to the island's children, as a large percent of them have visited, learned, and worked down at the cave with us.

ONE

Time's Most Important Moment

VISITORS COME TO HAWAII seeking paradise. But the truth is, these islands have become a kind of living hell for nature. The place is a microcosm of the world condition, where the role of humans in transforming nature stands out in high relief. This is a story that matters, because humans need to know that they are a threat to the rest of creation. Can we learn to tread more lightly on this speck of dust in the universe that we call home? Will anything we hold dear make it into the future? How did we get to this point?

For about 50,000 years *Homo sapiens* has been expanding its range. Starting out from the primordial cradle in Africa and Eurasia, humans have taken the rest of the world by storm, one landmass at a time (figure 1). We spread first to New Guinea and Australia, and ended with places never colonized prehistorically, such as the Mascarene and Galapagos Islands and Antarctica. It is fair to say that our species is native to Africa and Eurasia and a biological invasion everywhere else. We are an extremely successful weed. A global pattern of catastrophic extinction and landscape transformation tracks this human diaspora perfectly—too much so for coincidence. No use denying it: we humans can really pack a wallop, and have, time and time again, across the face of the planet.

This slow human wave has been washing over the globe for tens of thousands of years, and it has taken out Australia's giant extinct marsupials, the Serengeti-like large mammal diversity of the prehuman Americas, and even the giant lemurs that once lived on Madagascar.[1] But what do these phenomena have to do with the Hawaiian island of Kaua'i?

Kaua'i is one of those very remote islands reached late in the human expansion, perhaps as recently as a thousand years ago by most current

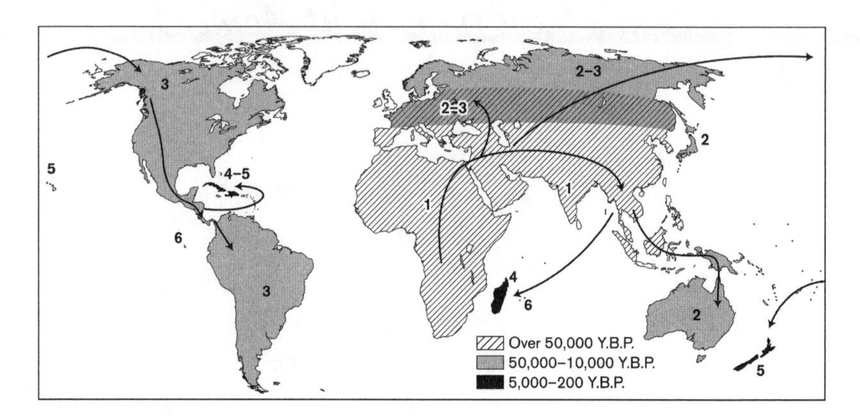

Figure 1. This map illustrates how scientists think humans spread across the planet and how their dispersal is associated with late prehistoric catastrophic extinction events. Modern humans arose in Africa and southern Asia (1), where large mammal extinctions were relatively light and mostly occurred more than 50,000 years ago. Roughly 50,000 years ago (2), humans spread to New Guinea and Australia. Meanwhile, they gradually moved northward in Asia after the last Ice Age (2–3), crossing over the Bering land bridge to the Americas (3) perhaps about 13,000 years ago. Although debate continues on the exact timing, large animal extinctions in the Australian and American regions seem to correlate at least approximately with human arrival. Humans then spread to large, moderately isolated island groups over the past 5,000 or so years (4), starting with Mediterranean and Caribbean islands, followed by Madagascar roughly 2,000 years ago. Again, extinction of the largest animals on these remote sites occurred about the time of human arrival. About 1,000 years ago (5), people reached Hawaii and New Zealand from South Pacific islands they had colonized earlier. Extinctions of large ground birds and other species followed. Finally, some of the most remote landmasses, such as the Mascarenes in the Indian Ocean and Galapagos in the Pacific, were apparently never colonized prehistorically, but the earliest written accounts detail the human toll on the dodo in Mauritius, the solitaire in Rodrigues, and giant tortoises there and in the Galapagos as well (6).

estimates. As a result, the wave of extinctions continues to break over isolated places like the Hawaiian Islands, and may be getting worse. Ever more species, now mostly plants and invertebrates but also some of the remaining few native vertebrates, are disappearing on our watch. Hawaii is the endangered species capital of the United States, with more listed than any other state—273 plants and 34 animals—all in only 0.2 percent of the nation's land area, one five-hundredth of the United States. If federally listed endangered plant species were apportioned by land area, Hawaii would be entitled to about one. If each state got an equal number, about 12. The official figures for Hawaii are a gross underestimate of the extinction challenge, as roughly *half the entire native flora* is considered at risk by most local experts.[2] Kaua'i, with a preponderance of single-island endemics (species found only on Kaua'i), is especially vulnerable.

I have been ruminating about this unique colonization moment in the history of any land—human advent and its subsequent consequences—for at least thirty years. I was studying the effects of human activities on cheetahs in the Masai Mara region of Kenya in the late 1970s for my master of science degree in conservation biology at the University of Nairobi.[3] Camped out for over a year in this northern end of the great Serengeti ecosystem, I kept asking my partner in this work, Lida Pigott Burney, the same simple question: *Why does Africa have all these wonderful big animals and other continents mostly don't?* It was an old question, we acknowledged, but with some recently proposed answers that were intriguing. This quest led us to the writings of Paul Martin, who has long been working on this topic, and getting scientists stirred up about it, with provocative articles starting with one in *Nature* over forty years ago titled "Africa and Pleistocene Overkill." In that article, Martin made a good first attempt at explaining precisely why Africa has so many big animals and other continents mostly don't: people and animals co-evolved in Africa and southern Eurasia; in most of the rest of the world, initial contact between arriving humans and a naive fauna was disastrous for the animals, because these technically advanced human hunters quickly overhunted any prey that didn't run away. This idea became known as "Pleistocene Overkill," and a mathematical model that illustrated how this could happen led to Martin's "Blitzkrieg Hypothesis."[4] This provocative idea about a human role in late prehistoric extinctions has spawned many others, and sparred with a few, particularly the climate-based and disease-based alternatives (figure 2).

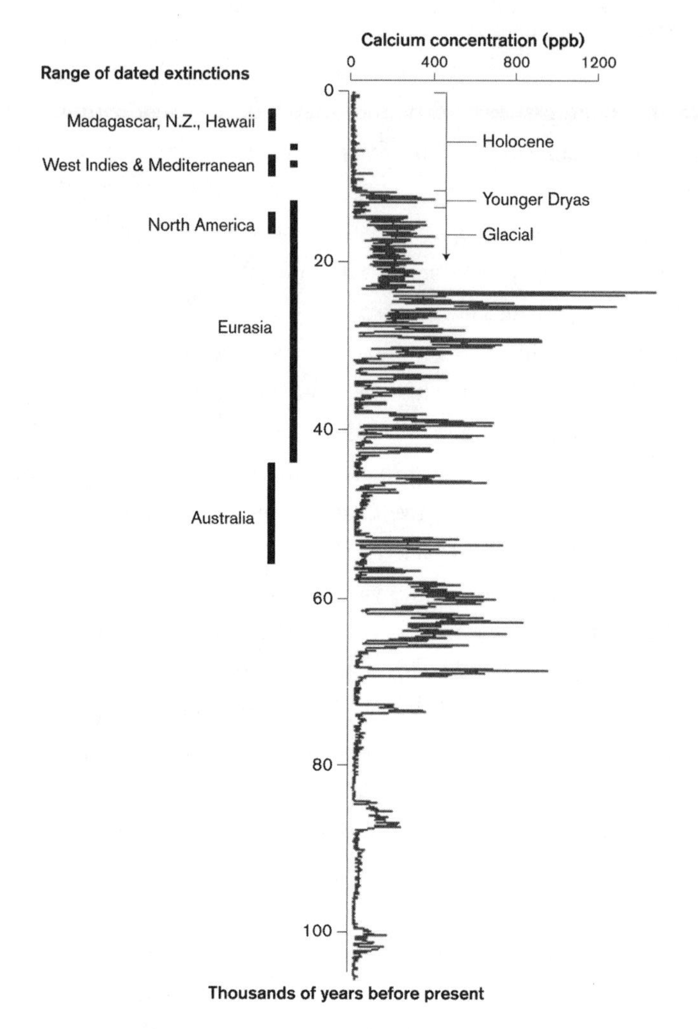

Figure 2. This plot shows the global trend in climate for the past 100,000 years as inferred from a high-resolution calcium concentration record from the GISP2 Greenland ice core. The calcium variation is caused by the relative amount of atmospheric dust, an index for cool and dry conditions (large values) versus warmer and wetter (much lower values). Comparison to the vertical bars, indicating the range of "last occurrence" dates for extinct faunas on various continents and islands, shows no correlation with climate change in general, occurring on different landmasses according to approximate time of human arrival, regardless of glacial or interglacial climate conditions.

Table 1. Hypotheses proposed to explain the late prehistoric extinctions

Hypothesis	Description
Climatic hypotheses	
Climate change	Climatic changes, in the form of a slow transition from mosaic vegetation to a more zonal pattern, lead to less hospitable environments for large herbivores[1, 2]
Rapid climate cooling	As above, but change develops rapidly at the Younger Dryas cooling event about 11,000 years ago[3]
Environmental insularity	Applied only to the extinction of the American mastodon; extinction occurs because boreal forest retreats northward after glaciation, resulting in expansion of deciduous forest, which is less hospitable to the species[4]
Overkill hypotheses	
Blitzkrieg, or rapid overkill	Animals were not afraid of humans and thereby rapidly hunted to extinction; extinction moving as a wave over landscape[5]
Protracted overkill	Overexploitation of initially naive fauna gradually leads to population collapse[6, 7]
Predator pit	Native predators contribute to a rapid collapse that is initiated by humans[8]
Second-order predation	A series of extinction events occurs as a result of interactions among humans, carnivores, herbivores, and vegetation[9]
Three-stage overkill	Rapid series of extinctions in which overkill alone is sufficient explanation[10]
Other hypotheses	
Clovis (paleoindian) age drought	Rapid onset of arid conditions leads to severe but temporary vegetation change following human arrival (Clovis culture), which amplifies the effect of human predation[11]

Table 1 (*continued*)

Hypervirulent disease	Infectious disease spreads rapidly through wide range of taxa, killing large animals differentially[12]
Keystone megaherbivores	Loss of megaherbivores (large browsing and grazing animals) that maintain open forest causes vegetation changes; for instance, fire regime changes as forests close and more fuel becomes available for burning[13–15]
Great fire	Landscape transformation by human-caused fire; extinction follows rapid loss of forage[16, 17]
Synergy	Human and natural causes interact; large herbivore decline leads to increased fire occurrence and landscape transformation[18]
Extraterrestrial impact	Comet or meteorite explodes over North America, producing instantaneous climate change and leading to large animal extinction in the Americas[19]

References

1. R. D. Guthrie, "Mosaics, allelochemicals, and nutrients: An ecological theory of Late Pleistocene megafaunal extinctions," in *Quaternary Extinctions: A Prehistoric Revolution,* ed. P. S. Martin and R. G. Klein (Tucson: University of Arizona Press, 1984), pp. 259–298.
2. R. W. Graham and E. L. Lundelius, "Coevolutionary disequilibrium and Pleistocene extinctions," in *Quaternary Extinctions: A Prehistoric Revolution,* ed. P. S. Martin and R. G. Klein (Tucson: University of Arizona Press, 1984), pp. 223–249.
3. W. H. Berger, "On the extinction of the mammoth: Science and myth," in *Controversies in Modern Geology,* ed. D. Muller et al. (New York: Academic Press, 1991), pp. 115–132.
4. J. E. King and J. A. Saunders, "Environmental insularity and the extinction of the American mastodon," in *Quaternary Extinctions: A Prehistoric Revolution,* ed. P. S. Martin and R. G. Klein (Tucson: University of Arizona Press, 1984), pp. 315–339.
5. P. S. Martin, "Prehistoric overkill: The global model," in *Quaternary Extinctions: A Prehistoric Revolution,* ed. P. S. Martin and R. G. Klein (Tucson: University of Arizona Press, 1984), pp. 354–403.
6. D. C. Fisher, "Extinction of proboscideans in North America," in *The Proboscidia: Evolution and Paleoecology of Elephants and Their Relatives,* ed. J. Shoshani and P. Tassy (New York: Oxford University Press, 1997), pp. 296–432.
7. S. L. Whittington and B. Dyke, "Simulating overkill: Experiments with the Mosimann and Martin model," in *Quaternary Extinctions: A Prehistoric Revolution,* ed. P. S. Martin and R. G. Klein (Tucson: University of Arizona Press, 1984), pp. 451–465.
8. D. H. Janzen, "The Pleistocene hunters had help," *American Naturalist* 121 (1983): 598–599.

9. E. Whitney-Smith, "Late Pleistocene extinctions through second-order predation," in *The Settlement of the American Continents: A Multidisciplinary Approach to Human Biogeography,* ed. C. M. Barton et al. (Tucson: University Arizona Press, 2004), p. 177.
10. J. Alroy, "A multispecies overkill simulation of the end-Pleistocene megafaunal mass extinction," *Science* 292 (2001): 1893–1896.
11. C. V. Haynes, "Geoarchaeological and paleohydrological evidence for a Clovis-age drought in North America and its bearing on extinction," *Quaternary Research* 35 (1991): 438–450.
12. R. D. E. MacPhee and P. A. Marx, "The 40,000-year plague," in *Natural Change and Human Impact in Madagascar,* ed. S. Goodman and B. Patterson (Washington, D.C.: Smithsonian Institution Press, 1997), pp. 169–217.
13. S. A. Zimov et al., "Steppe-tundra transition: A herbivore-driven biome shift at the end of the Pleistocene," *American Naturalist* 146 (1995): 765–794.
14. N. Owen-Smith, "Pleistocene extinctions: The pivotal role of megaherbivores," *Paleobiology* 13 (1987): 351–362.
15. W. Schüle, "Landscapes and climate in prehistory: Interactions of wildlife, man, and fire," in *Fire in the Tropical Biota,* ed. J. G. Goldhammer (New York: Springer-Verlag, 1990), pp. 273–318.
16. H. Humbert, "La destruction d'une flore insulare par le feu," *Mémoires Académie Malgache* 5 (1927): 1–80.
17. G. H. Miller et al., "Pleistocene extinction of *Genyornis newtoni:* Human impact on Australian megafauna," *Science* 283 (1999): 205–208.
18. D. A. Burney, "Rates, patterns, and processes of landscape transformation and extinction in Madagascar," in *Extinctions in Near Time: Causes, Contexts, and Consequences,* ed. R. D. E. MacPhee (New York: Kluwer Academic/Plenum, 1999), pp. 145–164.
19. R. B. Firestone et al., "Evidence for an extraterrestrial impact 12,900 years ago that contributed to the megafaunal extinctions and the Younger Dryas cooling," *Proceedings of the National Academy of Sciences, USA* 104 (2007): 16016–16021.

Next I was reading everything I could find on *paleoecology,* the study of past environments, and thinking about *human paleoecology,* a truly obscure subtopic in those days (even now). Scientists were learning a lot about the late prehistoric past, it seemed, by looking at microscopic fossils in lake mud. This work, coupled with other arcane methods such as tree-ring analysis, where scientists assess climate by measuring the thickness of ancient tree rings, was telling us a lot about past climate variation. But of this already small group, there were just a scant few (people like Paul Martin) who had been asking whether this kind of information could tell us anything about other forms of past change involving ecological dynamics (extinction, for instance), or even the activities of prehistoric humans.

In the early 1980s, I went to Duke University to complete a Ph.D.

with Daniel Livingstone, a paleoecologist who specializes in past environments of Africa. I had by this time discovered that there was an exciting methodological world that might help address some questions I had about human paleoecology. Livingstone, a bearded, ruggedly handsome Nova Scotian who had spent most of his career at Duke, had developed a type of piston-coring device for lake and marsh sediments, not coincidentally known in most scientific circles as a Livingstone Sampler.[5] Dan had worked everywhere from Alaska to perhaps a dozen countries in Africa. He was also a cousin of the African explorer David Livingstone, so he good-naturedly put up with a lot of lame "Dr. Livingstone I presume?" jokes.

Although the work was rigorously scientific, what I learned in that lab also seemed like a kind of potent scientific sorcery. Very few laymen and not even that many scientists back then had given much thought to palynology, or the study of pollen and spores. Dr. Livingstone's device brought up "sediment cores," which consisted of mud in long, thin, intact columns from the bottom of even the deepest lakes. He then treated samples from these cores in the laboratory with all kinds of strong chemicals like acids, bases, detergents, and solvents, finally deriving from the tormented sediments a residue containing millions of fossils that could be seen only with a microscope. Depending on the caustic recipe used, one was left with plant pollen and spores to show what some past environment had growing on it, or fossil diatoms and other algae skeletons to show what the water chemistry was like, or microscopic charcoal particles to indicate wildfire occurrence, or other arcane clues to past environmental dynamics. Radiocarbon dating and other methods provided a timeline for changes recorded in these core samples.

These techniques were powerful tools, and I was keen to learn them. Here was something very high-tech, on the one hand, and very down-to-earth on the muddy other, the kind of thing TV has sought to capture with shows like *CSI.* Lida and I took to this mixture of fishing, mining, plumbing, and forensic detective work, and were soon out drilling holes in all sorts of low places, eventually not just in lakes and swamps but in crater bogs, prehistoric impoundments, and, most important for future work, caves. Followed by lots of white-coat lab work and challenging hours, actually months, at the microscope, this endeavor added up to my Ph.D. project. My dissertation, *Late Quaternary Environmental Dynamics of Madagascar,* was all about this critical moment in time on this mysterious minicontinent off the southeast coast of Africa. Madagascar is very much like its big continental neighbor but with a sharply contrasting human history. Humans evolved in Africa, and have been there ever since. The first humans on Madagascar may have come ashore as recently as 350 B.C.[6] I used the scientific sorcery I learned in Livingstone's lab to try to answer the questions everyone working in paleontology, archaeology, and ecology in Madagascar had been asking for many years: *When did people arrive, how did things change, and what did people do to provoke these changes?*

That was the beginning of my adventures in "time travel." My wife, Lida, and our children, Mara and Alec, have made a lot of this journey with me. In addition, I have been lucky enough to have some great multidisciplinary collaborations with scientists who know a lot more about many of the needed specialties than I do. From them I have learned some speleology, paleontology, and archaeology to go with my palynology. What has emerged has been a search to find the right sites to study the questions of human arrival and its consequences. Old methodologies have been retooled and combined, new methodologies have been concocted, and a host of researchers, volunteers, and visitors have contributed to whatever these initial musings have added up to as a project.

Through huge coincidences, this global search led us eventually to a place on the Hawaiian island of Kaua'i. This green gem in the mid-Pacific is often referred to as the Garden Island and generally is equated in the movies and in the public psyche with paradise itself. Studying extinction there, of all places, with so many endangered species unique to this one island of the Hawaiian Archipelago, one could easily fall into clichés about Paradise Lost. But Kaua'i is an especially sweet land, the kind of positive microcosm where one can seriously entertain the notion that some of that paradise is still around, and that it might be possible to regain a little more of it with pleasant effort.

TWO

Proverbial Tracks

IN 1987, I GAVE A LECTURE AT a symposium titled "Early Man in Island Environments" held on the mystical Mediterranean island of Sardinia, off the west coast of Italy and just south of Corsica. Although I was there to talk about my findings in the mud of Madagascar, I met scientists working on similar topics on islands around the world. The potential insights that seemed to be harvestable from this kind of multi-island interest in human paleoecology pointed to the need for more of this type of information from islands everywhere. Two regions that interested me were the West Indies and Hawaii. This further growth in my already geographically wide curiosity led to new island-scouting adventures in exotic places like Antigua, Puerto Rico, Maui, and Molokai. Some were successful explorations, such as the work from Puerto Rico showing human arrival at 5,300 years ago derived from analysis of microscopic charcoal particles; or the ten-millennium pollen, diatom, charcoal, and trace metal record from the very high Flat Top Bog on Haleakala Crater, Maui.[1]

One place Lida and I really wanted to know more about was Kaua'i, which is about 5 million years old and the oldest of the major Hawaiian islands. We knew from the detailed early work of our friends Storrs Olson and Helen James, avian paleontologists from the Smithsonian Institution, that a remarkable bird fauna went extinct there after people arrived.[2]

Storrs and Helen are pretty remarkable, too. They are among the leading pioneers of avian paleontology, the study of fossil bird bones. That's not a very big field, but it is extremely interesting, particularly in remote island settings like Hawaii, where birds evolved to fill in ecologically for the generally missing mammals. Although they have worked all over the world both together and individually, some of their most well known and original work has been in discovering, naming, and studying the remains of the birds that went extinct in the Hawaiian Islands in the last millennium or so.

Storrs and Helen are southerners, like us—he from northern Florida and she from Arkansas. Storrs has the look and accent of an average southerner until you listen closely to what he is saying. He can be talking perfectly seriously in that Deep South drawl about the shape of the tarso-metatarsal bones of strigiforms (owls) versus raptors (hawks) or something equally dry and abstruse to most people, then suddenly crack a very funny joke and double everybody up with laughter. He is one of those people who sees and furthers humor in almost any situation. He also has an incredible trick: he will hold his hands behind his back, and you place any unidentified bird bone in them. Without looking, he will turn the bone over and over in his hands, still behind his back, then say something like: "Well, it's a proximal fragment of a left femur of a *small duck,* you know . . . hmm . . . let's see, with that shaft thickness it's got to be a Koloa or a Laysan Duck *around here,* unless you're trying to trick me with a smallish domestic mallard excavated in a restaurant . . . ha, ha, ha . . ."

Storrs laughs a lot, at both his own jokes and everybody else's, and he sees something funny anywhere he looks, sometimes a little crudely or worse. Helen is quieter, more sparing and careful with her own subtle humor, and striking in appearance with wavy red hair and a countenance one previous author described as "fresh faced."[3] She met Storrs, an established Smithsonian scientist, when she was an undergraduate at the University of Arkansas majoring in anthropology and doing a summer internship. That was about three decades ago. They got married, did a lot of research both together and apart, co-authored some really landmark paleo-ornithological papers, and collaborated with many other scientists on a host of fascinating projects around the world. They also have raised a boy and a girl, Travis and Sydney. They split up a few years ago.

So it was with those two, plus their two children and our two, together ranging from three to eleven in age, that we were looking for fossils on

Figure 3a. David Burney coring in Makauwahi Cave in late August 1992, along with (left to right) Travis Olson, Mara Burney, Sydney Olson, and Alec Burney. (Photo by Lida Pigott Burney)

Figure 3b. Excavating a test pit in Makauwahi's North Cave, with Alec Burney, age three, Helen James, and Storrs Olson. (Photo by Lida Pigott Burney)

Kaua'i's south shore near Poipu in late August 1992 (figure 3). The dunes in this area, along the scenic coastline of Pa'a and Māhā'ulepū, were the place that had yielded so richly for Storrs Olson and Helen James in the late 1970s. What they had found, and we were subsequently to find a lot more of, was a remarkable flora and fauna far more diverse than anybody had envisioned previously.

We found a few more fossil bird bones and extinct land snails at their old sites, and checked out other potential sites, including lava tubes in the Koloa area. One day Helen and I, with three-year-old Alec on my back, were making our way up the beach looking for fossils along the sea cliffs near the one house on this stretch, the home of the venerable Adena Gillin. The plan was to rendezvous with Lida, Storrs, and the rest of the kids about a mile eastward along the beach where some large bird bones had recently been seen protruding from a rock. We would have continued on up the beach to meet them, doing our part of the coast survey and not suspecting what lay just inland from there, had I not commented that it was odd that so many human footprints led in and out of a small sandy path. It bore directly inland from there, through a big patch of milo trees. Our curiosity paid off, because investigating those tracks changed my life forever.

Starting up the path, we met a tourist couple coming the other way. When we asked what there was to see up there, they gushed with enthusiasm about "the Warriors' Cave." To us, the word *cave* evoked special magic, as we both had found so many great fossils in caves. One look at this cave a few minutes later told us this was going to be a place to be reckoned with. It was getting late, so we made a quick once-over of the small cave inside the waist-high triangular entrance. We circled the big flat-bottomed sinkhole, nearly an acre of "sunken garden" inside featuring a huge banyan tree whose canopy covered almost half the opening overhead, and noted the large cave on the south end (which later came to be known simply as the South Cave) that would merit more thorough exploration tomorrow (figure 4).

To our surprise, when we reported our discovery to Storrs, Lida, and the kids over dinner at our campsite at Salt Pond Beach Park, Storrs knew all about the place. He showed us in his notebook the entry from a previous trip in which he had explored the place with Dr. William K. "Pila" Kikuchi of Kaua'i Community College. They had concluded that, although it looked like it should have potential for our type of research, a thick layer of modern clay blankets the site to several feet deep, and it was probably not a

Figure 4. The huge banyan tree in Makauwahi Sinkhole as it appeared in 1990. (Photo by Ed Miner)

high priority for further investigation. I recounted several similar features I had excavated successfully in Madagascar and Puerto Rico, and advocated that, next day, we take a small soil sampling device called a bucket auger and "poke some holes" there to see what lay below.

By this time, I had learned two perhaps obvious but easily overlooked fundamentals in this human paleoecology business. There is nothing more important than finding the right sites to answer your questions. Without those sites, nothing more can be done beyond speculation. And there is no way to find the right sites without exhaustive searching *below the surface.* Actually, looking beneath the surface of things is good advice in general, even the kind you might hear at a commencement address.

In our first few minutes in the cave the next morning, Lida, Helen, and I, with some young assistants, had bored a small hole in the silty clay floor of the entrance to the South Cave, and I was holding shells of the extinct land snails *Amastrella, Leptachatina,* and *Orobophana* in my muddy palm. At about 47 inches (120 cm) below the surface, we were stopped by sand collapse from a layer well below the modern level. We decided to try again, this time near the entrance in the North Cave. This hole yielded more extinct snails, but also the skull of a "good bird," the endangered Hawaiian coot. Heck, I thought, if we can get a bird skull with the second 3-inch hole we make in here, how many millions of bird bones and other interesting things must there be in this place?

We were all quite excited by our new site (figure 5). Good finds had been scarce this season, and up to this point it had looked as if Kaua'i was not going to yield much that was new. But the site is everything. A fossil location or archaeological site is a kind of vessel, a giant receptacle for information that must stay there long enough to be of interest at some later time. As not only bird bones and snail shells but also seeds, fish bones, and fine-textured organic sediments came up from the core, we knew we had made a nice little peephole into the past. All agreed that the sinkhole and cave system must have been flooded throughout recent millennia, and that well-preserved materials from a long range of time have collected in this giant vessel. How many thousands of years, of course, we didn't know

Figure 5. Oblique aerial view, from the west, of the Makauwahi Sinkhole, a collapsed-cave feature. Areas on all sides of the sinkhole are underlain with cave passages. (Photo by Ellen Coulombe, Wings Over Kauai)

until we had obtained our first radiocarbon dates on the third core we took, one made with our personal copy of Dr. Livingstone's contraption that penetrated down over 16 feet (5 m), and still didn't reach the bottom.

We left the island a few days later, not sure what we had but certainly intrigued by this potential new window into a past that none of us knew much about. Not long after, on September 11, 1992, 'Iniki, a hurricane more vicious than any in Kauaians' memories—the most powerful to strike the Hawaiian Islands in recorded history—flattened this island and set back its economy by several years. "It was a collective near-death experience" for people on the island, said then mayor JoAnn Yukimura. Six people were killed, but the toll could have easily been far greater. A Category 4 on the Saffir-Simpson Hurricane Scale, this storm had estimated gusts of 160 miles per hour (258 kph) or greater. Structures on Poipu Beach just near our cave were inundated to a depth of 20 feet (6 m) or more by the storm surge, and some storm waves reached 35 feet (10.7 m).[4]

Our sinkhole, of course, was profoundly affected too. For the first time since perhaps Hurricane ‘Iwa a decade before, the ocean surged up the stream outside the cave and flooded the floor inside. The great banyan toppled over, filling the entire sinkhole with its great brushy mass and a trunk the size of a whale.

THREE

Constructing a "Poor Man's Time Machine"

STORRS, LIDA, AND I WERE ABLE to juggle schedules enough to get back to Kaua'i in February 1996. We had all been very busy on other projects literally half a world away. My intervening field seasons in Madagascar had been some of my best, yielding a lot of new bones of extinct giant lemurs and fossil pollen records reaching back 40,000 years or more in parts of the island not studied previously from a paleoecological perspective.

By this time, more than three years after Hurricane 'Iniki, the sinkhole on Kaua'i's south shore was a frightful tangle of vegetation, and the great hulk of the old fallen banyan was at the core of it. The place had been transformed into a huge rank thicket of aggressive exotic vines, weeds, and saplings. From the rim, it looked like a vast chaotic bowl of salad greens. *We're going to core and excavate this?* I thought.

There was also a tangle of ownerships, jurisdictions, and involvements that surrounded the cave property. We began by seeking permission from the legal owner, Grove Farm Company, and the traditional Hawaiian owner, LaFrance Kapaka-Arboleda. County, state, and federal officials needed to be in the loop, and have the necessary paperwork in their files.

We especially wanted to be sure our growing friendship with Pila Kikuchi could turn into a sound local collaboration, something I learned early on in this work is essential to success on the local level. One government official told us off the record that if Pila wanted us to work there, we would eventually get our permissions, and if he didn't, we wouldn't. Pila was somebody locally who, when it came to antiquities, carried a lot of weight. He was an easy guy to like. A natural-born teacher, he knew quite a lot about a huge array of things, and his interest in everything rubbed off on the whole community. His huge curiosity was matched only by his wonderfully wry humor. Almost anyone who has lived very long on Kaua'i has Pila stories, often in connection with a course he taught them at Kaua'i Community College that opened their eyes to the fascinating local culture and history. His Japanese grandparents had come to Hawaii in the plantation era, so he was a third-generation local. Since earning his Ph.D. in anthropology from the University of Arizona in the mid-1970s, he had lived on Kaua'i with his wife, Dolly, a native of Indiana majoring in German at the same institution. At KCC, he taught the anthropology and archaeology courses and ran the small Department of Math and Sciences. This affable man and his entire family were to become like family to us over the ensuing years.

So he must have wanted us to get our permissions, because we eventually did. But no matter how rosy the situation might look in hindsight, we had some worrisome moments, as permission to dig for any purpose in Hawaii is not a given—and there was the possibility that what we were asking to do would be seen by somebody as essentially the desecration of a cemetery. Luckily for us, Storrs had many useful contacts in the islands, and the highly respected work he and Helen had been doing there for years automatically got our foot into a lot of the necessary doors.

Whomever we might have to reassure, and whatever red tape we might have to process, we were strongly suspecting by this time that the place was worth it. Back in my paleoecology and conservation biology laboratory at Fordham University, where I had been an associate professor since 1989, work had shown that the sediments of the site were rich in pollen, spores, diatoms, charcoal particles, invertebrate shells, vertebrate bones, wood, seeds, leaves, and DNA. We also anticipated that there would be archaeological materials in the upper layers. Our preliminary radiocarbon dating on the materials from the 1992 trip had shown that the site contained more than 5,000 years' worth of sediment, and that distinct layers represented many phases of change, including changes associated with human arrival and subsequent local human events.

We also learned that there were human burials up in high remote parts of the cave, and knew that, if we wanted to actually excavate in the sinkhole on the next trip, we would need to talk with the Island Burial Council. LaFrance, the traditional owner of the cave, was also the chairperson of this important group that oversees the protection of Hawaiian burial sites under county, state, and federal guidelines. But for the time being we had everybody's blessing to drill a few more small holes. I had learned the hard way from mistakes made on similar projects elsewhere that it is important to get as much information from cores and exposures as possible *before any digging starts,* so as to dig in the most promising places and with the right techniques.

By the time we had finished analyzing Core 6, a more comprehensive picture of what lay below had come to light. Many more analyses and radiocarbon dates later, we were able to reconstruct what we believe has been happening there for the past 10,000 years. Meanwhile, we completed this short trip by scouting out other potential coring and excavation sites around the island. As always, our friend Reginald P. Gage II, an avid local shell collector and general naturalist with many decades on Kaua'i, was helpful in finding sites. Reg is a retired navy captain who later worked for the county tax office and now does appraisals part-time. He is especially good for the sites with his precious *Carelia,* a genus of large, handsome land snails endemic to Kaua'i, all extinct of course, like most other native snails. Reg probably has the island's most comprehensive shell collection and a rich knowledge of books and other matters to do with old Hawaii.

Until this point we had financed the incipient Kaua'i investigation through a grant from the National Oceanic and Atmospheric Administration for the study of Human Dimensions of Global Change, which I had received for comparative work in Madagascar, Puerto Rico, and Hawaii.[1] This project was addressing questions of human arrival and prehistoric impact in three widely separated locations with one thing in common—diverse island ecosystems in each case had apparently collided with late-arriving humans. The Smithsonian Institution was also supporting this Kaua'i research, because the Bird Division at the National Museum of Natural History, where Storrs and Helen worked, was interested in finding more bird bones and gathering relevant stratigraphic and paleoecological information.

I was looking forward to a sabbatical in the 1997–1998 school year under a Fordham faculty fellowship, and Kaua'i looked like the place to spend it.

A grant from the National Science Foundation made it all possible. We had an offer of housing from the director of the National Tropical Botanical Garden in Kalaheo, Dr. William Klein, who was so encouraging at this stage that it is conceivable we would not have taken steps to follow through without his support. Kaua'i Community College was offering student help through Dr. Pila Kikuchi, who was also going to assist with the archaeological investigation. Dr. Patrick Kirch, a noted expert on Pacific archaeology from the University of California, Berkeley, and his students were also going to participate.

So what were we going to do? I hoped, figuratively at least, to build a time machine. Nothing fancy, really, just the presumably right kinds of research in a spectacularly promising place. I had been rattling on one day with paleoecologist Paul Martin about this idea, using stuffy terms like "integrated site analysis" and "landscape-level paleoecology," when he quipped, "So Burney is building a kind of poor man's time machine." The term stuck, and it aptly describes our overall goal—to get the clearest view possible of events that happened long ago on a landscape that has subsequently changed almost beyond recognition, and to do it on a small budget.

What is needed for this work is a fossil site that is more than the usual "snapshot"—that moment frozen in time when remains of organisms are caught in a situation that prevents the normal decay and dissolution that claims and recycles 99.999 percent or more of all things that ever lived. To look at something like the brief geological interval when humans first arrive and colonize a place, what is really needed is a site that is more like a photo album, or even a movie, than a snapshot. Imagine an album of photos of a landscape in which pictures are always accumulating, and which can be riffled through to show a kind of time-lapse movie of ancient events. The best fossil sites for studying our questions about human arrival and its consequences are like this. The sinkhole area of Makauwahi Cave was one of the best examples of this type of site I had seen anywhere (figure 6).

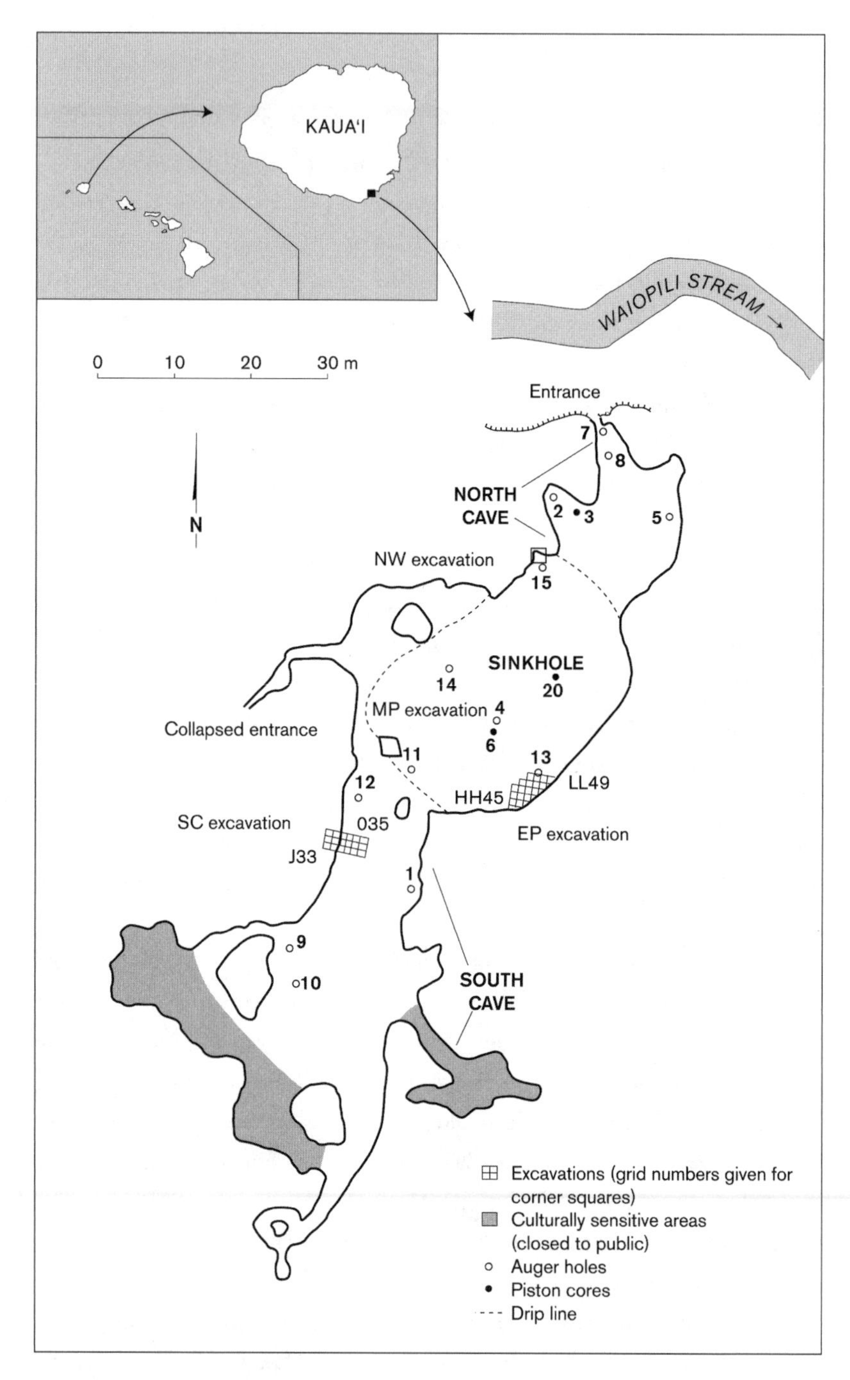

Figure 6. Map of the Makauwahi Cave system showing primary passages, excavation sites, and coring locations.

Finding such a place may sound like an impossible, pie-in-the-sky goal. But by this time we had already published studies like this for similar integrated sites and landscape paleoecology projects from Madagascar.[2] So we knew it could be done in Hawaii, too, if we could find the right sites, and that the story would potentially be useful for fleshing out the human paleoecology saga. The idea is to get a chronologically precise stratigraphic sequence, a great datable stack of sediments, fossils, and artifacts together, that give us the "before" and "after" of human arrival. We also hope we might get a good look at the "during" as well. With luck, we can distinguish in the record individual events related to the impact of the arrival of the first humans in this and other island contexts.

We have had some good results elsewhere from matching the palynology of a lake or marsh core with adjacent more conventional paleontological and archaeological sites, mostly through careful radiometric dating. This experience in Madagascar, Maui, and North America had showed us, however, that our best bet is in sediments that are under water and contain a wide variety of types of fossil evidence *together in the same site.* The great variety of sedimentary environments in Makauwahi Cave's many nooks and passages, and what was apparently an open lake in the sinkhole, provided many potential opportunities for our research.

The kinds of places needed for this work yield bones and shells of extinct and surviving animals and the pollen and seeds of the respective plants. The sediment is neither very acidic nor very alkaline—nearly neutral is best—and deprived of oxygen. This matrix of sediments should contain well-preserved microfossils such as pollen and spores. It also should ideally have a steady rate of new sediment coming in and not mixing much with older material. Trouble is, in practice that generally leads to working well below the water table, in sediments perfectly preserved in permanent groundwater. Excavating under such conditions requires a special technique for subphreatic, or below the water table, digging in permanently saturated mud while keeping straight what level things come from.

To accomplish this goal, I had been working on a small-scale version of what open-pit miners do. In the late 1980s, Lida and I first used this

Figure 7a. *The Exhumation of the Mastodon* (1806), a painting by Charles Willson Peale, depicts the first excavation of a mastodon at the beginning of the nineteenth century in Newburgh, New York. This is probably the earliest illustration of a subphreatic excavation, with a treadmill pump to control the water level and muddy but rewarding work for the excavators. (Used by permission of the Maryland Historical Society)

technique when we excavated the rims of some Carolina Bays.[3] These are mysterious little oval lakes and marshy basins that occur from Texas to New Jersey and reach their highest density in the coastal plain of the Carolinas. We did this work with Dan Bliley, a soil scientist from Smithfield, North Carolina. Dan is another example of a phenomenon typical of my entire career: I have worked with some of the funniest scientists around—experts with a sense of humor. Dan is another redneck like Storrs and me ("No, Burney," either one of them would probably say; "you ain't no redneck. From that scraggly beard I'd have to say you're most likely a damn southern hippie . . ."). We learned so much of a practical nature working with Dan,

Figure 7b. Across the Hudson River at Hyde Park, New York, investigators Dan Fisher (left) and David Burney (right) two centuries later use smaller and more efficient pumps to excavate another mastodon, but still get muddy. (Photo by Robin Andersen)

both about researching Carolina Bays, his hobby, and finding good sites for septic tanks, his regular employment. Dan was also working on his Ph.D. ("long term") from North Carolina State University, our undergraduate alma mater and the place where Lida and I first met late in 1969.

The basic subphreatic excavation method has been used in fossil sites since Thomas Jefferson's day. Jefferson witnessed this technique in the field, and the scene was immortalized by the artist and paleontologist Charles Willson Peale in a painting he made in 1806 (figure 7). As director of the Philadelphia Museum, Peale excavated a mastodon skeleton from subphreatic sediments at Newburgh, New York (not far from our home for fifteen years in Croton Falls). Jefferson was convinced that the beast was not terribly old, as the preservation was remarkable (of course—it was a Poor Man's Time Machine). Since this was long before radiocarbon dating could settle the age issue, Jefferson continued to believe these creatures were not extinct. His Deist religious beliefs, like most worldviews of the time, didn't admit the idea that anything in God's creation could be so imperfect as to go extinct. So he outfitted the explorers Meriwether Lewis and William Clark with the most potent light arms of the time, at least in part to prepare them for collecting a flesh-and-blood mastodon out West.[4]

In any case, as the seeming chaos of Peale's painting shows, one can excavate with relative comfort in flooded sediments by running a pump whose lift point is slightly lower than the level to be excavated. One then keeps the water level depressed by constant pumping while digging (creating a "cone of depression," in hydrological terms). What evolved from the inspiration of the likes of Thomas Jefferson and Dan Bliley was something that combines some of Dan Livingstone's coring concepts with regular archaeological techniques for excavating with three-dimensional recovery of information (*x*, *y*, and *z*-coordinates, in archaeological lingo). We drill a hole in a corner of the excavation pit, encase this hole with large-diameter PVC (white plastic) pipe so it won't collapse, then shove the suction head of a small gas-powered water pump down the hole. By adjusting the suction rate to keep the water at a constant level, one can employ the water itself as a leveling device. We then dig and screen the wet sediments in slabs usually 4 inches (10 cm) thick. One can moderate the pump speed to keep the water level within a centimeter or two, ideally, of the height of the lip of the casing pipe, which is set to 4 inches below the level of the top of the sediment to be dug off (figure 8).

It's great muddy fun, but potentially dangerous if one is careless about the engineering of the pit walls. Long stretches of soft, fully vertical wall can collapse without warning, which, if the excavation is large enough, could bury an unsuspecting individual alive. Big rocks hanging in the wall can succumb to gravity suddenly. Therefore, walls have to be terraced or slanted. This protocol drives most archaeologists crazy, because they love tidy, straight, square pits. Regardless of the engineering and esthetics, if we judiciously apply this combination of coring and excavating technology in the right sites, we can obtain a record of where fossils, artifacts,

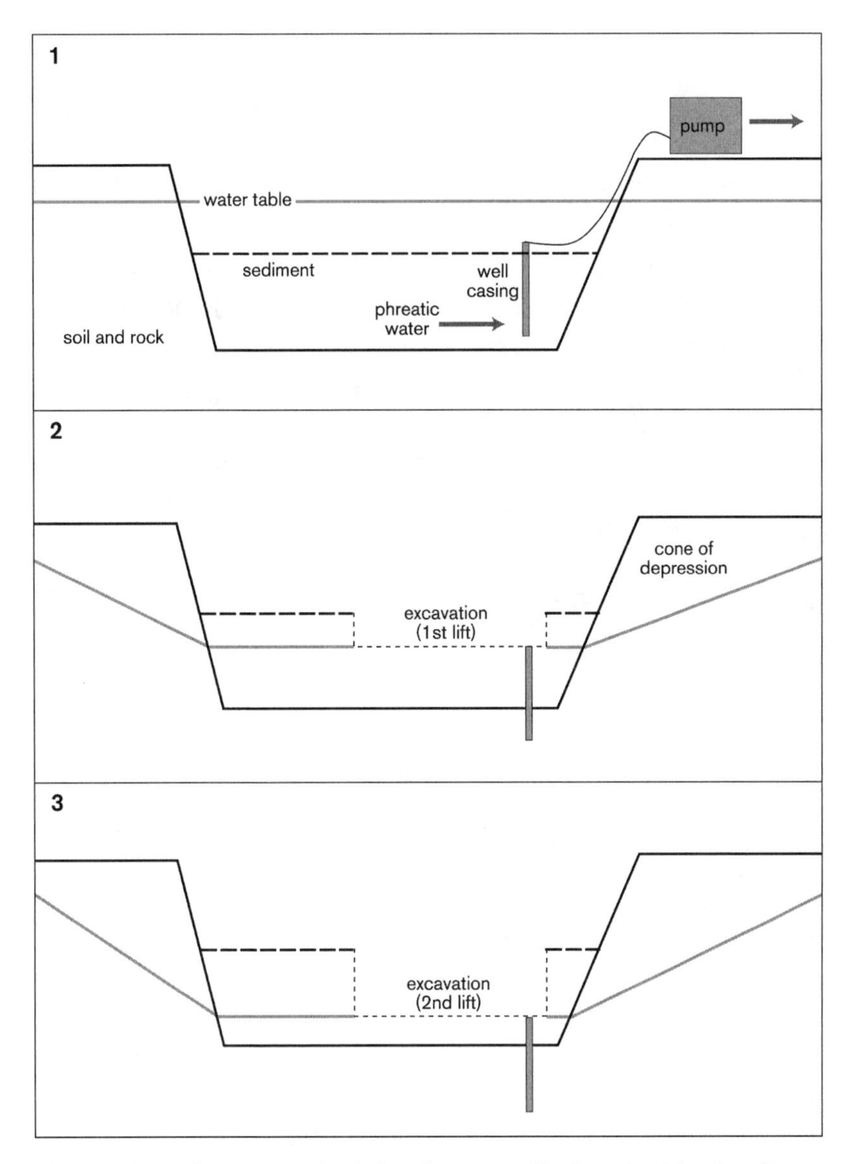

Figure 8. Stages in an excavation below the water table. In a pit with soft sediment in the bottom, we insert a large-diameter PVC pipe below the sediment surface and use it as a sump hole for the suction head of a pump. (1) We pump water to an adjacent pit and thus return it to the water table, but the removal of water creates a temporary water-table depression in the vicinity of the excavation. We then remove sediment to the level of the lip of the pipe, which is advanced in small, recorded increments to provide a sediment profile. (2) After we remove a small increment of sediment (generally 4 inches), we advance the pipe downward a similar distance and remove material to this new depth. (3) After we complete this level, we remove another increment in the same way, potentially many times over. I have excavated to as much as 16 feet below the water table with this method.

and sediment samples come from that is suitably specific for dating and stratigraphic analyses.

After many more cores, small test pits, and permission forms, we unofficially launched our time machine on September 7, 1997, when our group began to dig along the east wall of the sinkhole.

FOUR

Owl Omens

I am a brother to dragons, and a companion to owls.
—Job 30:29 (King James version)

I am a brother of jackals, and a companion to ostriches.
—Job 30:29 (Revised Standard version)

THE INITIAL RESULTS OF OUR first large-scale excavation along the east wall were mostly disappointing. Storrs and I had picked the place after coring all over the sinkhole and caves, but a big factor in choosing the exact spot to start digging was his opinion that owls would have roosted on the ledge above and dropped lots of prey remnants, and occasionally died and perhaps fallen into the mud below.

On the very first day of actual digging, Pila Kikuchi had mustered a huge force of KCC students and other interested folks in the community, and that meant that, once the leaves were raked up and the vegetation removed, there wasn't that much for them to do along the east side where Storrs and I had decided to dig. Excavation is an incredibly tedious thing. But these folks had mostly worked with Pila before, so they already knew it was going to be a slow slog at first, and they soon settled into working on some shallow test pits in the South Cave. It was clear from the cores I took right along the edge of the places we intended to dig that the twentieth century was a thick layer of silty clay with little of interest for 3 feet (1 m) or more down, unless you are a collector of 1950s memorabilia or something similar. We already knew the story of how so much 1950-to-the-present sediment got in there.

Elbert Gillin, a Grove Farm engineer, lived for decades with his wife Adena in the one house along the wild Māhāʻulepū coast. In addition to his monumental work creating some of Kauaʻi's water tunnels through mountains (to allow water transfer from one watershed to another, a popular, expensive, and sometimes successful irrigation trick during the island's plantation days), Mr. Gillin also drained the large brackish pond called

Kapunakea that stood outside the cave and extended up into the area now quarried away for limestone. He cut a ditch to the ocean in the early 1950s. The pond was largely drained off and the Waiopili Spring by the adjacent limestone quarry then fed into a small dredged channel. Just upstream from the cave, this channel joined a big ditch that straightened a former stream channel draining the cane fields of the central part of the Māhā'ulepū Valley. Today these freshwater sources pass by the entrance to the North Cave through a narrow man-made canal, then on out to the ocean west of the Gillin house. In other words, Elbert and his associates were quite willing to modify drainages on a big scale, a trademark of Hawaii's plantation era. As one might expect from the modern perspective, Elbert's scheme, along with the nearby limestone quarrying activity and large-scale farming in the vicinity, apparently contributed to massive periodic erosion and redeposition of adjacent soils. The cave and sinkhole deposits resulting from this erosion produced what we came later to refer to as Unit IXB of the excavations. Water pouring off the fields and quarry down Mr. Gillin's ditch during storms in subsequent years added more than a meter of sediment to all the interior of the cave system except the high sandy portions in the back of the South Cave.

So our dig in the sinkhole, for many days, yielded only a scattering of artifacts from a baby-boomer's childhood: cellophane, plastic, Styrofoam, aluminum pop-tops, Polaroid film backs, and so on. A few bird bones, from the introduced mynahs and doves, turned up along with far more bones of chickens, large rats, and cane toads. Certain snails introduced by humans marked certain dates in these young sediments; for instance, we know that the carnivorous rosy wolf snail (*Euglandina rosea*) was introduced in the mid-1950s by the Hawaii Territorial Department of Agriculture in an unsuccessful attempt to control other introduced snails. When the rosy wolf snail's shells were no longer present in the sediments, we knew that we had dug past the era when Chuck Berry, Ray Charles, and Elvis appeared.

This slow yield of the remnants of modern progress was really boring for most of the original volunteers and spectators. Day after day, though, Storrs and I dug on, continuing even after Helen and Lida had returned home for the season so the kids could start back to school. Most volunteers had long since satisfied their curiosity.

The next major layer, as the sediment changed abruptly from the dark gray, very sticky clay that was just below the 1950s layer, was dune sand. In the field microscope I could see that the grains of this sand were almost perfectly rounded. These tiny limestone and silica grains shaped like pills and ping-pong balls are the same kind of sand that makes up the modern dunes along the coast near the sinkhole. This material contained the kinds of things one would expect for the late nineteenth and early twentieth centuries: a horse skeleton, teeth and bones of goats and cows, a piece of sail canvas, bits of iron—but mostly just tons of sand. We removed the sand from our pits and made great piles of it that to this day provide useful landscaping materials in the sinkhole.

A picture that Pila Kikuchi found in the Bishop Museum Archives, believed to date between 1890 and 1920, tells the tale. A view of the landscape showing Kapunakea Pond, the limestone cliff with the entrance to the North Cave, and the adjacent dunes and fields was almost as stark as the landscapes beamed back by Martian soft landings. Bare soil, rock, and sand as far as one can see. No trees. A bush or herb here and there, mostly leafless. Very little grass.

This was a landscape eaten alive by feral livestock. Old stories confirm that a primary use of the area for decades was as free range for livestock. Hunting feral animals was a pastime and sometimes a pest-control challenge. During this time on the adjacent island of Ni'ihau, for instance, Aubrey Robinson and his friends, relatives, and workers killed thousands of goats. To this day, Ni'ihau hosts a sport-hunting operation that seeks to control feral pigs and mouflon sheep. In 1917, Aubrey apparently succeeded in his goal of eliminating goats from his smaller island, something Kaua'i and the other major Hawaiian islands have most certainly not been able to do. Ironically, there are many goat hunters who are the primary advocates for not wiping them out, as well as being some of the few folks with the hunting skills to actually do the job. They understandably would prefer to hunt goats in perpetuity rather than have one massacre. They also know how hard it is to get those last few in the rugged terrain of the interior of an island like Kaua'i.

Meanwhile, back at the excavation—at the bottom of a thick layer of nineteenth-century sand, in finer sands banded with clay and silt, we finally came to a level full of goat teeth, and then, just below, no more iron or Eurasian feral animals. We had passed on below Captain Cook's time. By now we were finding a few fishhooks of pearl shell and bone. But the good stuff was sparse in this pit, even after weeks of digging. Although I was intrigued by the fact that these late parts of the island's story were represented, we were beginning to wonder if the key earlier layers were going to be out of our reach. By this time, we were also beginning to do the pumping and digging routine necessary to get below the water table and still maintain proper archaeological decorum—that is, know the depth and location of artifacts and other interesting features in terms of *x, y,* and *z* coordinates.

Just as we were getting to fairly rich if somewhat muddy archaeology, as well as some nicely preserved seeds of plants introduced by the Polynesians, we hit those awful rocks that had stopped our corer in places we had tried along the eastern wall. This was the stuff of prisoners' nightmares. Hundred-pound stones (even bigger in later parts of the dig) formed a kind of jagged slippery pavement several feet thick on top of what we hoped would be rich early Polynesian and eventually prehuman layers. Most of these were not smooth limestone boulders. They were a highly angular, rough, mostly unweathered chaos of blocks and slabs of basalt, lithified (hardened to stone) red soil, coral rag, and the cave's own limestone, tightly packed in a matrix of gravel, sand, clay, and wood. Later study revealed that this was a remarkable tsunami deposit from about four centuries ago. This punishing layer was not only tiring and time-consuming to get through but also was in no way amenable to "textbook" excavation. I do a lot of my digging in rich deposits with just a small trowel and scoop, and often with just my bare hands in soft mud full of artifacts, bones, and wood, so as not to scratch or otherwise damage the evidence. But this tsunami deposit was the kind of place where we had to employ picks, shovels, sledgehammers, and chisels or go home. And I was planning to stay awhile, armed with a sabbatical, good grant funding, free housing, and the only slightly diminished belief that this site was going to be one of the best we ever worked in a long career of this kind of craziness on the islands of the Atlantic, Pacific, and Indian Oceans.

Storrs and I were really discouraged, though. My paleoecological work can use almost any kind of data that comes along, as the first order of business is to try to infer the past environment, whatever it was. But Storrs and Helen needed bird bones to stay engaged with this effort, and that is what we promised the National Science Foundation they would find. They are very accomplished experts in a small important scientific niche—avian paleontology—so I knew that other commitments would soon crowd this project out if it didn't start yielding something of more than passing avian interest.

So there we were, one afternoon in late September, digging a bit, screening, picking out what little there was of interest, digging some more, moving some rocks, screening, picking, digging, moving rocks, digging, moving rocks, screening, on and on like this. Standing down in the pit, we looked at the pavement of big rocks still below us waiting for painstaking removal, reflected that we couldn't be more than a few hundred years back in time and still were apparently a long way from any prehuman extinct bird deposits—and despairing.

Half-jokingly, I suggested to Storrs that it was time to try my old buddy Dan Bliley's "P.H.D. method," as he called it. Dan would take, in a difficult juncture like this in our Carolina Bay excavations of a decade earlier, what he called a P.H.D. (that's a post-hole digger, a tool well known to almost anyone who ever built a fence by hand). Dan would make a sounding, dropping each chunk brought up by the pincers on a separate labeled screen or basket. By measuring the hole depth after each plunge of the P.H.D., one gets a rough idea of stratigraphic depth. Unlike a coring device, the wide jaws of this ancient double-shovel contraption grab enough stuff to get a pretty good sample of what is below if one screens each morsel separately. With the advanced P.H.D. skills of a Dan Bliley, you can wedge and wiggle your implement's way between and around buried obstacles.

We generally frowned upon such a "quick-and-dirty" method, but sometimes it's either take desperate measures or give up. We promised each other that we would only do this enough to satisfy ourselves that we hadn't wasted several weeks of our lives on a "dry hole." Once we had any kind of sign from below that there really was a trove beneath that we could reach—remember, all our optimism about birds was based on that one coot skull we found in Core 2 five years before—we would stop this plunderous approach and get back to removing the overburden with duly slow diligence.

So down we went, racing back in time in our incipient Poor Man's Time Machine by launching a P.H.D. probe. Within minutes we had jumped back a millennium or more, lifting out bones of large birds and seeing no more evidence that humans were still in the picture.

We congratulated ourselves that the past had sent us a powerful omen when I held in my hand a moment later the remains of one of our favorite of the island's extinct bestiary, the long-legged Kaua'i owl (*Grallistrix auceps*). Owls are regarded as special, even otherworldly creatures in most traditional cultures, and they certainly figure prominently in native Hawaiian lore. Hardly any owl could be more profoundly mysterious than this unusual extinct one (figure 9). Its long legs and pointed wings presumably were perfectly suited for catching avian prey on the wing (an odd occupation for an owl, but *Grallistrix* lived in the essentially mammal-free world of prehuman Hawaii). As students of classical mythology as well as Hawaiian lore, it was easy for Storrs and me to see this as richly ironic and well timed, a kind of signal from the denizens of that lost world brought by one of their most auspicious dead messengers, saying that perhaps we should stick around and keep digging a while longer.

We cleaned, bagged, and labeled our day's catch, then headed out. Just as our rental car was turning from Mrs. Gillin's driveway onto the main Grove Farm road, we were startled by the fluttering wings of a big owl. It was *Pueo,* the surviving native Hawaiian short-eared owl (*Asio flammeus*). The creature hovered momentarily just above our windshield. It wheeled overhead, and came back to hover again ("as if to be sure we got the message," I wrote in my notebook that night). Then it made a heart-stopping

Figure 9. In a prehistoric scene from the cave vicinity created by the paleontologist and artist Dr. Julian Hume, the extinct Kaua'i owl *Grallistrix auceps* scatters a flock of the extinct finch *Chloridops wahi* as it pursues the more recently extinct Kaua'i o'o (*Moho braccatus*). (Monochrome of color painting by Julian Pender Hume)

plunge toward the ground just alongside the car, and caught a mouse. The owl settled on the ground farther away, still in plain view, to enjoy this rodent tidbit.

Storrs and I were uncharacteristically quiet on the way back to Koloa. Whether we had witnessed a genuine signal of some kind, or just saw an owl do a spectacularly beautiful series of maneuvers, almost didn't matter. We were absolutely sure, from the bony evidence we had collected, that our luck had changed for the better. An owl from past times, never to be seen in life again, and his magnificent living cousin, had each welcomed us whether they meant to or not. We split at least one bottle of champagne that night. The Poor Man's Time Machine had landed squarely in the prehuman past of Kaua'i.

FIVE

Opening Ancient Doors

OVER THE NEXT THREE YEARS and more at this big sinkhole we dug three large pits, moving more than 260 cubic yards (200 cubic meters) of sediment. Mud from layers with fossils and artifacts was washed through a 1/16-inch (1.5 mm) mesh of window screen (figure 10). Thousands of bones, shells, and seeds and hundreds of human artifacts were bagged, labeled, stored, and studied. More than three hundred volunteers—students, local people, and tourists—put at least a day or two, some many weeks or even months, into combing through the labeled buckets of this wonderful mud I carried up the ladder from the pit every day. Along the east wall, we made it to the limestone floor of the cave at 21 feet (6.3 m) below the surface. Sediments were even thicker near the middle of the sinkhole, reaching back 33 feet (10 m) and 10,000 years. The pit in the South Cave yielded many wonderful artifacts, including perishable items of wood, fiber, and gourd.

By the summer of 1998, we had opened up a lot of ancient trapdoors that led down into a little-known time before humans on Kaua'i. It often seemed literally like this, as we pried up a large stone, and there below it were the soft muds of a millennium ago, and older times were archived just under that. Beneath the great rain of rocks from a tsunami of about four centuries ago lay the early Polynesian period, a rich layer of lake sediments containing bones of pigs, dogs, chickens, and the small Pacific rats that came with the early Polynesian colonists. Remains of plants brought by the early Hawaiians were much in evidence, as were perishable cultural materials like objects of wood, bamboo, fiber, and gourd, and evidence for past human diets.

As the sediments grew more fibrous and organic below, all traces of

Figure 10. Screening sediments at Makauwahi Cave in 1997. Alec Burney, age nine, looks on as Dr. William K. "Pila" Kikuchi makes a good find.

humans disappeared, and there we were, looking into the wall of a pit at the time immediately before, during, and after the first humans arrived on Kaua'i. Bingo! On and on we dug, ultimately picking out bones of more than 40 species of native birds, a dozen species of extinct land snails, and the seeds and wood of the plants that grew nearby for the thousands of years immediately preceding human arrival. Bones of birds and shells of other creatures previously unknown to science were included in the rich smorgasbord of owl, hawk, waterfowl, wading bird, and sea-bird bones, as well as the tiny bones of honeycreepers and other small perching birds once found on Kaua'i. About half the bird species represented are extinct, at least on Kaua'i, and many of the rest are endangered. Here we have discovered several new species, including two extinct honeycreepers, *Rhodacanthis forfex* and *Loxioides kikuchi.*[1]

The plant fossils held a big surprise. We recovered not only the seeds, pollen, and wood of the plants typically associated with the surviving remnants of native vegetation in the area, but also many others, including extremely rare plants that occur today in only a few isolated pockets in the high interior. We were intrigued to see that some plants that are very rare today, even virtually extinct, were thriving on the site for thousands of years before the big changes of the past millennium.

At the bottom of this wonderful rich prehuman layer—6,000 years and as much as 16 feet (5 m) thick in some parts of the sinkhole—we found quite something else. Seven millennia ago, the cave was breached by the sea, turning the place into a giant subterranean tide pool for long enough to accumulate a layer of beautifully preserved marine mussels and other seashells. During this time, areas of the sinkhole even out near the center appear to have been roofed over. Perhaps the ocean churned the cave's interior, and created huge blowholes like the ones that exist nearby today at Spouting Horn and other south shore locations. In any case, something caused the roof to collapse at about that time.

Beneath this marine layer mixed with roof collapse, we were back in a kind of reddish-brown clay, typical of the material deposited on dry or seasonally flooded cave floors throughout the tropics. Below this, we were stopped by rock, the apparent limestone floor of the cave. This great pile of mostly soft sediments 20 to 33 feet thick (6 to 10 m) on the floor seemed like an edifice in time, with windows we were beginning to pry open at many different floors. But the Poor Man's Time Machine cast a wider net still, as our group explored more cave passages and the deposits in the quarry next door. Looking at the complex layering and channeling, we were able to extend this story, with less precision of course, back nearly a half million years.

The whole trick in doing paleoecology is to *think like a landscape.* Aldo Leopold pioneered the idea of thinking like a mountain. Paul Martin wrote about thinking like the Grand Canyon.[2] In this case, we were striving to think like a cave on Kaua'i.

More than 500,000 years ago, the Māhā'ulepū Valley was sculpted from the lavas extruded during Kaua'i's last fit of volcanic eruptions. These volcanoes spattered along a big crack in the older rocks of the island (up to 5 million years old). This major fault formed as the aging island began to break apart and gradually subside into the sea.[3] The new coastline at Māhā'ulepū, like all of the world's Pleistocene tropical seashores, moved from offshore to inland and back again on an Ice Age climate cycle of roughly 100,000 years. During each Ice Age glacial maximum, sea level stands much lower than today, often roughly 400 feet (120 m) lower, or even more. As a warmer period between ice ages, or interglacial, developed, sea level rose with melting glacial ice, to temporarily raise levels to approximately the present height for 10 millennia or more; then the world slipped back into an Ice Age and sea level dropped again, stranding the high interglacial beaches.

Based on radiometric dating of volcanic layers that overlie the cave limestone in the adjacent quarry, geologists Chuck Blay and Paul Hearty have independently estimated that the layer of sandy limestone the cave has formed in was deposited on the beach here as dune sands roughly 400,000 years ago.[4] As sea level rose quickly after the glaciers melted, the ocean waves advanced inland, grinding coral and coralline algal reefs into limey sand. This sand piled up on the beach, and the prevailing trade winds from the east drove the sand grains inland and heaped them into great sand mountains like the modern ones at Barking Sands, on the island's southwest coast.

These dunes, often piled on top of older, solidified dunes from an earlier interglacial, are eventually colonized by terrestrial vegetation. They form topsoils, grow trees, sometimes get dusted by volcanic ash or scorched by new lava, and generally become mature landscapes. In the process, rainwater leaches down through the sand mountain, carrying terrestrial organic acids and salts derived from volcanic soils and the marine environment. Over the eons, the sand dune itself becomes a fossil. It petrifies. Aragonite, a fairly soluble form of calcium carbonate, dissolves from the deposit and recrystallizes into a harder, less soluble calcite. The silica from volcanic sources also helps harden the mix. The resulting wind-deposited "natural cement" is properly known as eolianite or eolian calcarenite, in reference to Aeolus, Greek god of the winds. Inside the sinkhole the eolianite preserves the original striped fabric of the wind-blown sand layers, often set at a tilt that reflects the sloping profile of the original dune, never more than about 30 degrees. Otherwise, the sand would have exceeded its "angle of repose" and slid off the side of the dune. Even sand grains obey quite strict rules of behavior.

Today in several parts of the cave you can see distinct coarse eolianite sand layers, laid down by strong winds, interbedded with thin lines of finer sand that reflect more moderate winds. One can easily imagine each stripe as a day's worth of wind, as it kicks up in the late morning, moves some big sand grains along, then begins to die down and winnow out the finer grains toward sundown.

This great fossil dune field probably sat around not doing much but growing trees and occasionally getting bashed by high interglacial sea levels for perhaps a couple of hundred thousand years. In any case, things were stable long enough for the rock that would later form the cave to become quite hard. This limey sandstone or sandy limestone is laced with veins of white, yellow, or greenish calcite that would have formed after the rock solidified, settled, and cracked.

Some of these cracks, near the base of the layer, served as conduits for the abundant groundwater that pours down to this coast like a great subterranean sheet of freshwater. Many people are surprised to hear that this usually dry leeward coast of the island has vast amounts of freshwater beneath. But remember, some of the wettest mountains on earth are directly inland, uphill from the site. Mount Waiʻaleʻale at the center of the island is widely cited as the wettest, second wettest, or one of the wettest spots on earth, with as much as 40 feet (12.2 m) of rain per year.[5]

This underground water carries plenty of carbon dioxide in the form of carbonic acid, as well as other acids derived from organic matter and volcanic rocks. Flowing under our fossil dune field, this water slowly undoes the natural cementation process, eating away at the eolianite's lower layers. By perhaps 100,000 years ago or more, an extensive cave had been carved from the rock, reaching something like its present configuration at least 10,000 years ago. But this is a kind of limestone, so some of the water dripping and flowing from the roof and walls of the almost entirely enclosed cave environment caused the recrystallization of calcium carbonate and other minerals, creating a host of beautiful and bizarre flowstone and dripstone features known as speleothems. These wonderful formations—stalactites, stalagmites, straws, helictites, "cave popcorn," flowstone curtains—first awed me as a small child touring Luray Caverns in Virginia with my parents, a half day's ride from our native High Point, North Carolina. But where else in Hawaii did anyone ever see such a collection of speleothems?

Some of these are really big, 20 inches (0.5 m) or more in diameter. Similar-sized flowstone features that I worked on from Zaire, Botswana, Somalia, and Madagascar in the 1990s, which our group dated by the uranium series technique as well as radiocarbon, took tens of thousands of years to grow.[6] They persisted on cave ceilings, walls, and floors for well over 100,000 years. So when I walk through the cave showing the decorated passages to visitors, I can't resist reminding them that they are seeing a lot of time in one sweep—10,000 years of soft sediment deposited on the floor of the cave, walls decorated with dripstone that may be ten times as old as the sediment, mantled on rock five times as old as that. *Time vertigo* is what I call this strange feeling of seeing so many time scales at once.

We are talking about a half million years of time here as if we can really imagine such great spans. In any case, the dripping darkness of the cave was rudely interrupted about 7,000 years ago, as the sea found its way into the cave's passages when sea level rose following deglaciation. The central passage, the cave's biggest, must have been almost fully enclosed for a long time before collapsing into the sinkhole, as many large speleothems mantle its dry walls. They are not forming today, but they would have been able to grow in the 100 percent humidity of the great now-collapsed passage, the kind of environment required for deposition of calcium carbonate from dripping water.

The radiocarbon-dated sediments of this time show the dramatic change, as the dusty floor of the cave was overwashed by the rising sea, leaving beach sand and marine shells behind. But ocean waves do some dramatic things in the confines of a cave. The thin, high ceiling of the central passage collapsed during this event. Some of it may have already come down during earthquakes or earlier marine events. One can easily imagine huge blowholes and deep churning tide pools at this stage. Today large blowholes emerge from sea-cave roofs just down the road a couple of miles at Spouting Horn. In my travels I have seen two sinkholes very similar to ours, but in this slightly earlier stage of marine modification. These are at Pancake Rocks, on the South Island of New Zealand, and Andrahomana Cave, near Madagascar's southern tip. Ocean incursion can remake a cave in a geologically short time, as in all three of these places. But at Māhā'ulepū, something happened that was very convenient for our purposes. After the roof collapsed, a great deal of rock and sand apparently also closed off a passage of the South Cave that led toward the ocean. The result of the changes, and perhaps a small recession of sea level, was that freshwater (remember that subterranean freshwater cut the cave in the first place) formed a pond in the caves and particularly in the newly formed collapsed passage—our beloved sinkhole (figure 11).

For the next nearly seven thousand years, right up until the time of Elbert Gillin in the early 1950s, this remarkable place was not just a cave, but also a lake. For paleoecologists, this is two ideal fossil-forming environments for the price of one. Especially during dry weather, this great subterranean pool must have served as an irresistible water source, and potential deathtrap, for any terrestrial creature that came by. The system was also tenuously connected to the aquatic and marine environments nearby, through a narrow passage from the North Cave to Kapunakea Pond on the outside. The present walk-in entrance to the cave linked the pond inside with the larger pond on the outside, a great sheet of brackish water that was probably connected to the ocean over a tidal bar, like many other unchannelized streams in these islands. The thick organic sediments of the sinkhole lake are thus full of bones of marine fish like the mullet or 'ama'ama (*Mugil cephalus*), which likes to swim up streams. They probably swam into Kapunakea on a high tide, and some unlucky few ventured through the cave entrance, almost certainly never to find their way out. The diatom flora of the cave, analyzed by a Fordham graduate student I advised, the

Figure 11. A teenage Alec Burney enjoying a zip-line view of the Makauwahi Sinkhole.

late Dierdre McCloskey, showed that the waters of the sinkhole pond were fresh to slightly brackish. So the Pleistocene dry cave turned churning tide pool turned freshwater lake went on, millennium after millennium, attracting living creatures and entombing them in this giant pickling jar. This wonderful mud was full of not only the bones of birds and the shells of extinct land snails, but literally bushels of seeds and big chunks of unpetrified wood. Microfossils of pollen, spores, diatoms, charcoal—even ancient DNA fragments—were well preserved, too.[7]

Thin bands of coarse marine sand and masses of wood—even whole tree trunks—probably attest to storm overwash from hurricanes or other extreme marine events thousands of years ago. About 4,000 years ago, an increase in pollen from dry-adapted plants and microscopic charcoal particles suggests perhaps a century or more of extreme drought with "natural" (that is, prehuman) fires. But by and large, the place just sat there for all those thousands of years, soaking up sediments and fossils from the humanless living landscape.

About a thousand years ago, a big change occurs. People arrived on Kaua'i by this time, their presence loudly announced in the cave sediments by the first appearance of bones of the Pacific rat (*Rattus exulans*). Apparently this smallish rat came to the island via double-hulled Polynesian voyaging canoes (perhaps as an invited guest, because they were a culturally accepted food item). Bones of other nonhuman passengers (dogs, pigs, and chickens) also begin to show up in the sediments. Polynesian plant passengers also turn up, including kukui nut (*Aleurites moluccana*), gourd (*Lagenaria siceraria*), and small coconuts (*Cocos nucifera*). Microscopic charcoal particles drastically increase in the sediments, heralding the fires ignited by the invading bipedal primates. Large flightless birds, bird-catching owls, and the larger snail species disappear at about the time of the first human evidence, and some of the common types of seeds drop out, including the endemic *loulu* or fan palms (*Pritchardia*).

One day four centuries ago, something pretty extreme happened here. Huge rocks rained in over the east wall of the sinkhole, undoubtedly carried by a very large tsunami. To lift stones weighing as much as 220 pounds

(100 kg) from the beach and carry them into the sinkhole over a wall about 27 feet (8 m) above sea level, as this tsunami did, must have required waves that were much higher than that, with the great strength of a mega-tsunami. This event really stirred up the sediments, blurring the record from early Polynesian times in some deposits. We learned from cores and excavations all over the cave that the brunt of the extreme marine deposit (huge stones of basalt, red lithified soil, coral rag, and the cave's own eolianite) was thickest along the low, ocean-fronting southeast rim. From there the tsunami deposit thins out toward the back of the two caves, producing fans of sandy detritus-laden sediments.

There must have been a Hawaiian village along the coast on the outside then, as there was again in the early nineteenth century. Within this tsunami deposit are the sorts of things a coastal village would have. The layer contains pieces of canoes and paddles, ropes, tool handles, whole gourds (smashed flat, of course), and ornamental objects.[8]

After the tsunami, the lake was much shallower, perhaps even shoaling in the South Cave and along the east wall. People moved back into the area (but we are fairly sure they never lived in the cave, because it was too wet). Their food items, fishhooks, and debris of ordinary life accumulated on top of the ragged tsunami rocks. Fewer and fewer fossils of native organisms turn up, although many native birds, and some of the snails and plants, are still in the vicinity at this time around three centuries ago.

At a layer that contains more erosional clay than before, we see the first evidence for Eurasian contact: abundant goat teeth and bits of iron. In ensuing decades, almost all traces of native plants and animals disappear from the site, and they are replaced by remains of feral livestock such as sheep, horses, and cattle. These introduced herbivores ate up all the vegetation. Denuded of plants and soil-holding roots, the reactivated sand dunes then began a slow march, through the late nineteenth and early twentieth centuries, moving along downwind and dribbling over the limestone wall into the sinkhole, where even sand cannot easily escape. Erosion by water speeds up the process of filling the sinkhole in the mid-twentieth century, as Elbert Gillin channelizes, the quarry next door gets quarried, the cane

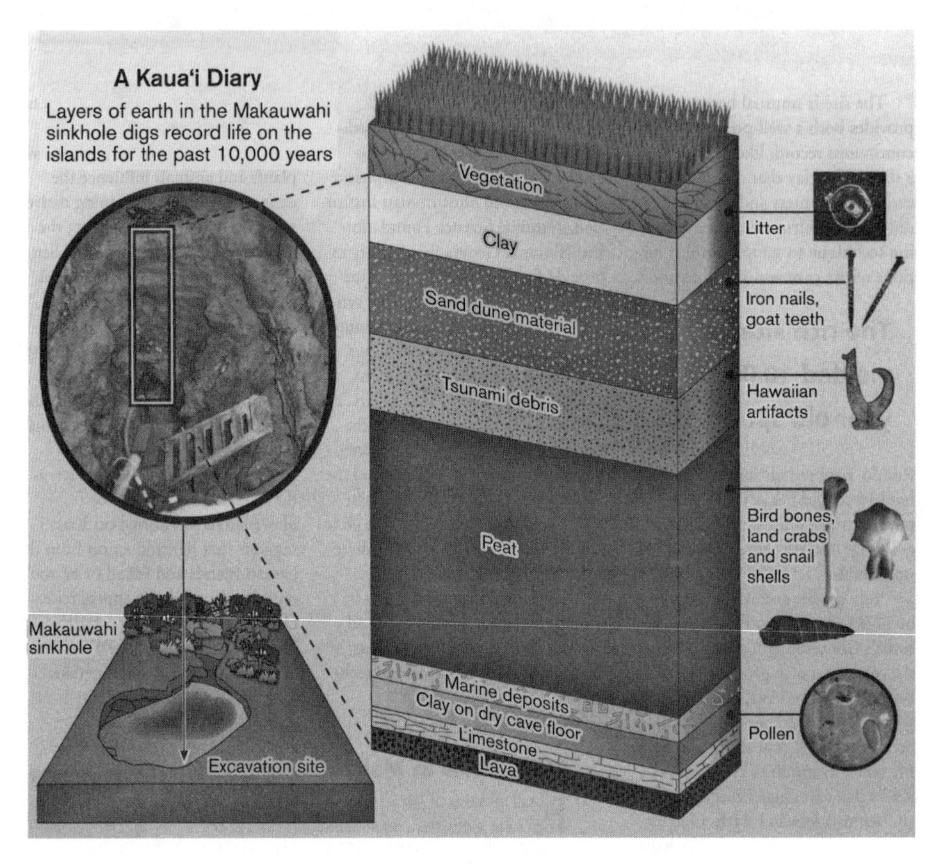

Figure 12. A cutaway diagram of a 10,000-year slice of sediments, nearly 33 feet thick, near the center of the Makauwahi Sinkhole. (Illustration by Martha Hernandez for Malamalama, University of Hawaii)

fields erode, and next thing we know, we are standing in the sinkhole on a dry silty clay surface that covers a meter or more of Styrofoam-bearing deposits (figure 12).

So there's the site history in brief, as best we know it. My colleagues and I have published this evidence and lots of analyses in more technical detail elsewhere, of course. It's a nice story outline to have, but to this point in the narrative we are lacking many details, especially concerning these big changes of the early human era.

SIX

Characters and a Stage, but No Script

WORK GROUND ON AT THE CAVE, month after month, through the rest of 1997 and into 1998. We dug some huge pits, and found a lot of fossils and artifacts. There's no denying that. But it was not a perfect situation. The work was incredibly strenuous, particularly in the huge and ever more huge pit along the east wall. This East Pit, as it came to be called, has been the main place yielding the vast quantities of bones of extinct birds. By early 1998, I was facing the challenge of keeping this ever-widening and deepening chasm engineered in such a way that it would not collapse on me. The cave wall adjacent to the pit needed careful checking for new cracks every day as work commenced, in the unlikely chance that our undermining of the sinkhole walls was weakening this ancient structure in some way.

As a veteran of these kinds of creepy pits, I was not terribly worried, as our engineering seemed sound, with a wide margin of safety. We all agreed that the safest approach was to keep a backup pump handy all the time, wear a helmet, and minimize traffic in and out of the pit. This boiled down to a lot of work for me, not so much to dig the sediment, which was generally quite soft below the rocky tsunami layer, but to carry the buckets, labeled with depth and grid location and weighing generally 65 pounds (30 kg) each or more, up the ever-lengthening extension ladder out of the pit. We decided this was much safer than hoisting the buckets up, as that would have carried the risk of bumping the walls and dislodging rocks onto my head. A helmet is not much protection against a 20-pound (10 kg) rock falling several feet.

I like to tell people, when they balk at donning a helmet before coming down into one of the pits for inspection, photography, or whatever, about the time my dirty old Petzl caving helmet probably saved my life. I was down in the East Pit one evening, working late, about 16 feet (5 m) below the surface. The local feral chickens were getting restless to have their sinkhole back, because there were still several large Java plum trees inside that they used for roosting at night. The dominant rooster of this clan was standing on the rim of the sinkhole, roughly 65 feet (20 m) directly above me. He was scratching his feet, displaying his impatience for the changing of the guard. Apparently he dislodged a small rock, roughly thumb-sized. Falling that short distance, it gained enough momentum to strike the unsuspecting top of my head, making a cracking sound on the helmet like a rifle shot. My head was jerked back forcefully, leaving me with a sore neck for several days but otherwise unharmed.

So I tell people, when I help them buckle the helmet, that one might need to wear this thing for a lifetime of this kind of work to be sure it's on that one time when it is needed. Some events are incredibly rare but can have terrible enough consequences to justify extraordinary precautionary measures. But aside from being pleasantly tired and occasionally a little sore from hauling 30 or 40 of those muddy buckets up a long ladder each day, I worked steadily. This site was yielding too richly to even contemplate taking time for much of anything else for many months. And I just love digging, whether I'm finding a lot or not.

By the summer of 1998, we were beginning to find some time to work a few other sites around the island. There were lots of good reasons to do so. Most important was that we had learned enough from our digs in the sinkhole by now to start formulating some ideas about when and how the area had changed so much since people came. The problem is, it is always inadvisable to formulate a history for an entire landmass, no matter how small, from one site, no matter how rich.

Another reason we started addressing some other sites was that we had slipped into a phase in which we had been forced to adopt a nomadic lifestyle when on Kaua'i. The housing provided by the National Tropical Botanical Garden had evaporated. Bill Klein, the director who offered us this nice collaborative arrangement, died suddenly from a heart attack just as our project was getting under way. NTBG was many months getting a new director, so we were allowed to stay on in the interim. A month after a new director was hired, however, in February 1998, we were evicted. Dr. Paul Alan Cox seemed to be mostly focused on his own specialty, medical ethnobotany, and apparently had no interest in the sort of thing we were doing. His secretary asked us to move out, because he had other plans for the space—"accommodating movie crews." We protested that this was part of the agreement with the National Science Foundation, with the invitation letter from Dr. Klein appended to the funded proposal, and that we would have budgeted for housing otherwise. There seemed no possibility for compromise. We had no choices but to move on or muddy the waters, so we moved on.

For the rest of 1998 and in subsequent years, as we continued during holidays and summers after I had returned to Fordham following the sabbatical, we house-sat for people, camped, took short-term rentals, and generally made do as best we could. This put a strain on us, in terms of hard digging six or seven days a week coupled with portable housekeeping, but in hindsight it did serve a purpose: since the County of Kaua'i puts a limit on how long one can stay in a given campground (maximum of six days), we fell into a pattern, particularly in the summer of 1999 and for the next few years' field seasons, of camping all over the island. We would spend a few days per month scouting out, coring, and excavating near these campsites all over Kaua'i. The rest of the time we devoted to digging at the sinkhole.

These other sites, I hoped, would fill in some of the gaps in our story about prehistoric Kaua'i and its subsequent changes. As we began analyzing and publishing the cave's voluminous results, it became more and more apparent that we now had a long list of characters, and were defining the "stage" pretty well, in terms of our reconstructions of what the vegetation and other basic parameters of the local landscape looked like, but had not really nailed down a "script." By that I mean that many of the key action elements of the story, and the critical timing of certain events, were still vague. Here is a list of what we really needed to know more about to have any hope of accurately reconstructing past events:

1) When did people actually arrive on the island, and where exactly did they come from?
2) What did they do on the landscape?
3) What did these people eat?
4) When did the various species that disappear from the cave's record actually die out?

It should be clear from the nature of these questions that we needed not just time traveling, but also to visit a range of locations in the past. Information from the wetter, windward coast of the island would be useful to compare with our leeward, dry site. We wanted to look at events in the higher parts of the island's interior. And finally, we should look at other types of sites, so that we could be sure that our conclusions were not being shaped by peculiar biases of this particular site.

This is the approach that we have been calling landscape-level paleo-ecology, or just landscape paleoecology. In Madagascar, we had used 23 sites, scattered all over the island, to address these kinds of questions and have added some others in subsequent years. My graduate student Guy Robinson was meanwhile getting a good start on similar questions in up-state New York, with five sites. These are other long stories, but the take-home message for our work on Kaua'i and elsewhere was that this approach can take us pretty far in answering some of the key questions listed above. In Madagascar we had established human arrival with multiple criteria (first appearance of pollen from plants introduced by the original colonists, a sudden drastic increase in microscopic charcoal particles in sediments, an increase in pollen of plants associated with deforestation, and earliest dates on human-modified bones of extinct animals). These all pointed to human arrival slightly more than two millennia ago on the Big Red Island.[1]

Studies of the Malagasy language of Madagascar in the 1950s had already shown a separation from its closest surviving linguistic relative, Maanyan, in the highlands of Borneo (yes, Borneo, in Southeast Asia) about the same time. So these remarkable folks who became the Malagasy ap-parently came not from adjacent Africa, but all the way across the Indian

Table 2. Clues that paleoecologists use to determine when humans have arrived in a region they previously did not inhabit

Method of detection	*Rationale*	*Examples*
Bones of introduced microvertebrates	Because of high reproductive rates, small animals brought by humans will be highly visible in the stratigraphic record soon after human arrival	Pacific rat (*Rattus exulans*) in Hawaii[1]
Increase in weed pollen and spore types	Vegetation disturbance by humans leads to large increase in weedy pioneer species	Increase in bracken fern spores after Maori colonization in New Zealand[2]
Appearance of exotic types of pollen	Plant brought by first colonists might naturalize and produce a distinctive pollen horizon	Appearance of *Cannabis* pollen in Madagascar[3]
Fossil traces of cultural water pollution	Arriving humans might transform watersheds, releasing nutrients to water bodies that change the fossil plankton flora	Increase of nutrient-loving algae after local human settlement in western Madagascar[4]
Sudden increase of microscopic charcoal particles	Use of fire by humans leads to an increase in soot particles above normal values	Detection of human arrival in Australia from microscopic charcoal in Lynch's Crater[5]
Drastic decline in dung-fungus spores	*Sporormiella* spp. grows primarily on megafaunal dung; thus distinctive fungal spores might provide a proxy for large mammal density; sudden decline might indicate transformation of mammal biota by humans	Spore decline before Younger Dryas cooling event and charcoal increase in upstate New York[6]

Table 2 (*continued*)

References

1. H. F. James et al., "Radiocarbon dates on bones of extinct birds from Hawaii," *Proceedings of the National Academy of Sciences, USA* 84 (1987): 2350–2354.
2. M. S. McGlone and J. M. Wilmshurst, "Dating initial Maori environmental impact in New Zealand," *Quaternary International* 59 (1999): 5–16.
3. D. A. Burney et al., "A chronology for late prehistoric Madagascar," *Journal of Human Evolution* 47 (2004): 25–63.
4. K. Matsumoto and D. A. Burney, "Late Holocene environments at Lake Mitsinjo, northwestern Madagascar," *Holocene* 4 (1994): 16–24.
5. C. S. M. Turney et al., "Redating the onset of burning at Lynch's Crater (North Queensland): Implications for human settlement in Australia," *Journal of Quaternary Science* 16 (2001): 767–771.
6. G. S. Robinson, L. P. Burney, and D. A. Burney, "Landscape paleoecology and megafaunal extinction in southeastern New York State," *Ecological Monographs* 75 (2005): 295–315.

Ocean, from somewhere in South Asia. This is the same general region and language family (Austronesian) that gave rise to the Polynesians who eventually reached Kaua'i. How ironic that, although separated by literally half the globe and forty degrees of latitude, from one subtropical zone to the other, the people of these two places we study may have common ancestry a few thousand years ago. These first and greatest of the ancient mariners had shrunken the globe long before the Europeans awakened from their medieval slumbers to "discover" the rest of the world. As Jared Diamond has pointed out, finding people with Indonesian language and culture just off the African coast is exactly as if Columbus had been greeted in the West Indies by blond-haired Norse-speaking Vikings. Come to think of it, if the Viking colonists in Greenland and Newfoundland had been a little more clever about learning from their Native American neighbors, something like Madagascar could have happened in the Americas.[2]

We also were able in Madagascar to establish some other key time horizons. Our dating of virtually the entire extinct large animal fauna showed that most of these creatures survived for a thousand years or more after these proto-Malagasy came ashore. By comparing all the types of data outlined, we were able to suggest that it has taken the Malagasy many centuries to conquer their island minicontinent.[3] Perhaps the many African and Asian diseases there, particularly malaria, dysentery, and plague, have played a role, partially slowing down human population growth at least until the twentieth century. What really would have helped, though, would have been some way to establish not just when species went extinct, but when their populations began to decline.

We found a tool that seems to fill this purpose. *Sporormiella* is a genus of coprophilous fungi that produces abundant, distinctive spores that preserve as a microfossil in the same muds that contain pollen, microscopic charcoal particles, and so forth. *Coprophilous* is a great word for fungi or other organisms that grow on and actually feed on dung. Owen Davis did pioneering work with *Sporormiella* in western North America in the mid-1980s, showing that it is common in some lake sediments until about 12,000 years ago, when the mammoths and three-quarters of the other types of large mammals disappeared from the continent.[4] It reappears, he found, after Europeans introduced livestock to the West.

This tool worked splendidly for us in Madagascar, showing at multiple sites that the fungus spore, and presumably the big mammals that produced most of its food, declined sharply within a few centuries of human arrival.[5] Only after that, we noted, did a sharp increase in charcoal particles actually manifest itself, suggesting that the loss of the big grazers and browsers (pygmy hippos, giant tortoises, elephant birds, and giant lemurs) might have triggered widespread fire and vegetation change. As plant litter accumulates on a previously heavily eaten vegetative landscape, modern studies show, fire is likely to increase during dry seasons, with a loss of fire-intolerant woody vegetation. People may thus have changed Madagascar and perhaps other previously uninhabited lands by first hunting out the bigger animals. This action set in motion a whole series of events, such as increased burning, vegetation change, and perhaps human dietary stress in the face of the ensuing shortage of large prey, that could interact synergistically. These factors, a kind of "extinction vortex," to borrow a term from Michael Soulé and other conservation biology pioneers, could also interact with factors inherent in the system but not generally lethal before these changes, such as natural climate variation, demographic limitations imposed by very small populations, and extinction cascades, in which the decline of one species may lead to the decline of other species somehow interdependent with the first.[6]

A similar story, with the lowly dung fungus spores as the centerpiece, also has emerged from the work in New York state of my former graduate student Dr. Guy Robinson.[7] These two studies were relatively big news in the media for a while after they came out in the journals *Science, Proceedings of the National Academy of Sciences,* and *Ecological Monographs.* Dozens of reporters for newspapers, magazines, and broadcast media contacted us. One actually asked me if this dung-spore thing was for real or just some kind of scatological joke. All this attention led me to quip on one occasion that "even a lowly fungus growing on crap has its Warholian 15 minutes of fame."

It would take years of work to find out whether a similar approach, looking at pollen, spores, microscopic charcoal, and animal bones, would also work on Kaua'i, we realized. So to hedge our bets, while we have continued to work the "proven reserves" in the cave, we have devoted a portion of our research time on Kaua'i since 1998 to finding and analyzing new sites for these key indicators—a charcoal spike, appearance of exotic introduced pollen types, evidence for deforestation and increased weediness, and latest dated occurrences of extinct species and first dated appearance of introduced animals. If *Sporormiella* or something like it could work here, all the better, but we knew not to hold out much hope since the dung piles here would be from large flightless ducks and geese, not mammals. Bird dung is chemically too different, we suspected, to appeal to the same coprophilous fungi. It was almost certain, though, that the combined methods we had used in Madagascar would tell us something interesting about Kaua'i if we just took the trouble to look for the right clues all over the island (figure 13).

One of the complicating factors in understanding our results from the cave and sinkhole was its coastal position. The system is adjacent to a coastal estuary that has been altered repeatedly by hurricanes, tsunamis—and people. We wanted to see if the kind of marine-driven dynamism (including prehistoric mega-tsunamis) that the record from the cave seemingly showed was typical of other coastal estuaries. Another uncertainty in our overall

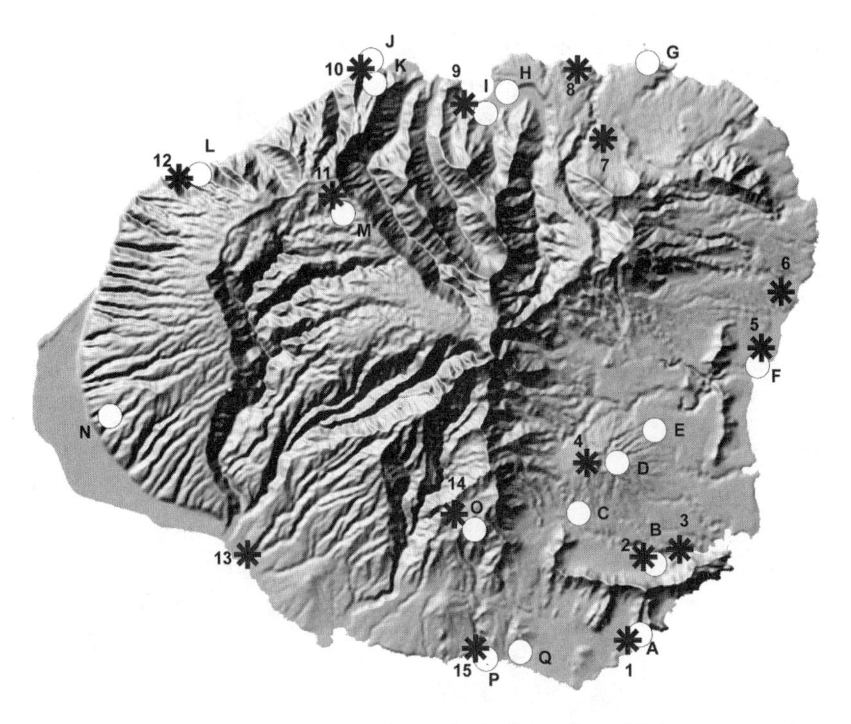

Figure 13. Map of Kaua'i showing paleoecological sites (asterisks) and restoration projects that have drawn from paleoecological results for guidance.

Paleoecology sites: (1) Makauwahi Cave; (2) Huleia Stream; (3) Alekoko (Menehune) Fishpond; (4) Kilohana Crater; (5) Kawaihau Wetland; (6) Kealia Beach; (7) Silver Falls; (8) Anini Beach; (9) Waipa Fishpond; (10) Limahuli Bog; (11) Alaka'i Swamp; (12) Nualolo Kai; (13) Kekupua Fishpond; (14) Kanaele (Wahiawa) Bog; (15) Lawai-kai Fishpond.

Restoration sites: (A) Makauwahi Cave Reserve; (B) Huleia National Wildlife Refuge; (C) Kahili Mountain Park; (D) Iliahi (Grove Farm); (E) Kapaia Reservoir (Grove Farm); (F) Kawaihau Wetland Restoration (Ducks Unlimited); (G) Kilauea Point National Wildlife Refuge; (H) Hanalei River Restoration; (I) Waipa Cooperative; (J) Limahuli Lower Valley Preserve (NTBG); (K) Limahuli Upper Valley Preserve (NTBG); (L) Nualolo Kai State Park; (M) Alaka'i Swamp Wilderness Area; (N) Pacific Missile Range Facility (U.S. Navy); (O) Kanaele Bog (Nature Conservancy); (P) Lawai-kai and Lawai Valley Restorations (NTBG); (Q) Koloa Caves and Shearwater Colony (Kukuiula Development Corporation). (Map courtesy of Jonathan Carbone) picture was how Polynesians had affected the coastal story with their remarkable engineering feats, particularly the fishponds they constructed. Pila Kikuchi of KCC was a leading authority on these, and he had been the one to point out in the journal *Science* that native Hawaiians might well have been the world's first aquaculturists.[8] We wanted to know when the ponds were constructed, and how the landscape changed during this period of Polynesian occupation of the island before written history.

We also wanted to find more sites that would tell us about the extinct land snails, a little-explored topic on Kaua'i. The record of stepwise disappearance of these snail species from the cave's sediments (first demonstrated by my daughter Mara in an award-winning high school science project) suggests that the extinctions actually may have occurred as a series of three discrete events. And finally, we wanted to see what was going on up *mauka,* or inland—that is, we needed to look at sediments of bogs in the interior to see what was happening at low, medium, and high elevations.

SEVEN

Fishponds

ON THE SOUTH COAST OF KAUA'I, about 5 miles (9 km) west of the cave, is Lāwa'i-kai, a place of legendary beauty and the former home of National Tropical Botanical Garden founders Robert and John Allerton. Geologically and hydrologically, it is typical of small estuaries in the Hawaiian Islands. The stream's headwaters drain a steep cleft in the basalt highlands, arriving at the coast with waters that under normal circumstances are clear and fast-flowing. Several times per year, and particularly during hurricanes and winter's *kona* downpours, the little stream "flashes" and delivers great volumes of muddy brown water to the shore. Lawai Stream is tenuously and intermittently impounded by a longshore sandbar at its mouth. This flat area flooded with brackish water drains through the beach sand both below the surface and through ephemeral surface channels. On very high tides and during stream flooding conditions, the sandbar may breach to considerable depth, or even disappear altogether, opening the estuary to marine influences and also draining it rapidly on the next falling tide.

On the west side of the channel, extending through a broad tidal marsh, there is a large sediment-filled prehistoric fishpond of the type Pila Kikuchi classified as *loko kuapā,* a natural coastal estuary that Hawaiians impounded with stone walls and packed earth (figure 14). With the help of a graduate paleobiology field class from Fordham University and local volunteers, I cored this estuary and the fishpond in the summer of 1998. We recovered in the middle of the ancient fishpond our longest record from this site, a core 21 feet (6.5 m) long that penetrated sediments well below the level of the man-made fishpond. Radiocarbon dating indicated that the

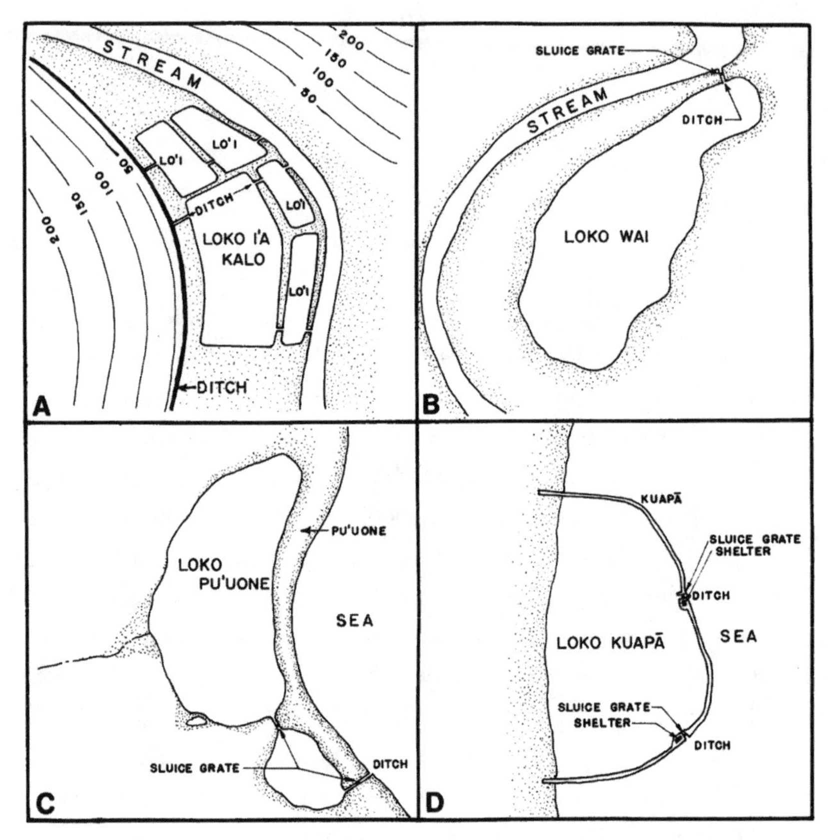

Figure 14. Pila Kikuchi's drawings from three decades ago illustrate the four basic types of Hawaiian fishponds: (A) ponds dug in inland areas primarily for taro cultivation; (B) a natural lake artificially connected to a stream by a ditch; (C) ponds created by coastal barrier beaches; and (D) a seawall isolating a coastal body of water. (From W. K. Kikuchi, "Prehistoric Hawaiian fishponds," *Science* 193 (1976): 295–299; used by permission)

base was at least 6,000 years old. More recent sediments associated with the fishpond period were hard to interpret in terms of age, because the original pond bottom seems to have been dredged down to older sediments. Subsequently, it was choked right to the surface by marine overwash, from a couple of tsunamis and Hurricanes 'Iwa and 'Iniki.

The prehuman sediments were lovely, though, if you like mud as

Figure 15. Alekoko Fishpond near Lihue, Kaua'i, a remarkable example of Hawaiian stonework that was probably built more than five centuries ago. It is often called Menehune Fishpond because of the widespread legend that it was built overnight by the magical menehune people.

much as I do. The bands of sand, silty peat, peaty sand, sand with snails, sand/peat laminae (fine layers), and dense organic clays gave way at the bottom to sand with marine shells. This stratigraphy was remarkably similar to the cave's sequence, and even contained many of the same pollen and snail types.[1] Lower Lawai Valley was flooded by the rising postglacial sea about six millennia ago. This was followed, as at the cave, by several thousand years of brackish to freshwater ponding, occasionally interrupted by extreme marine or flood events. We don't know how ancient the ancient fishpond actually is, because of the way the bottom appears to have been disturbed by dredging and subsequent marine overwash.

But east of the cave, on Kaua'i's southeast corner near the county seat of Lihu'e, we thought we might learn more from what is probably the most famous ancient fishpond of all—properly Alekoko, but known to locals and tourists alike as the Menehune Fishpond (figure 15).

A registered National Historical Landmark, it is a place of great traditional, archaeological, and historical interest as one of the finest examples in the entire archipelago of prehistoric stonework and fishpond construction. Said to have been constructed overnight by magical ancient people called *menehune,* the pond was formed by walling off a meander on the north side of the Huleia River. The great underwater wall runs for over 1,900 feet (600 m), apparently made from a dirt base faced with fitted stones, some with flattened sides. In places, the wall has been patched subsequently with bags of concrete, iron roofing, and wooden stakes. This probably is the source of the alternative story that the pond was built by Chinese plantation workers in the nineteenth century, although the late Francis Ching, a local archaeologist of Chinese ancestry, demonstrated from old photographs and historical records that the pond predates the arrival of the first Chinese laborers.[2]

Siltation and encroachment by introduced mangroves and other exotic vegetation has considerably reduced the area of open water in the venerable pond from its original 39.5 acres (16 ha). At low tide the pond is mostly less than 3 feet (1 m) deep, except near the two breaks in the wall. Perhaps these were locations in the old days for the *makahā,* gates said to have been made from mountain-apple wood (*Syzygium malaccense*) with slats below the water level that allowed tidal currents and small fish to pass unimpeded, but prevented the escape of larger fish. Anyway, I cored this pond to nearly 13 feet (4 m) by myself from a platform of poles cut from the wood of nearby *hau* trees (*Hibiscus tiliaceus*), with two inflatable kayaks for floatation. Most coring efforts are rugged team efforts at best, but I was alone and unaided this particular day in January 1999. With Christmas vacation over, the rest of the family had returned by then to New York, and my local friends were all busy with post-holiday activities. It was a strenuous day, but I got the core, showing that the estuary had been laying down riverine sediment there for over three thousand years. A big change comes in the sediments, to finer clay and more organic matter, at a layer full of woody debris that dated to 580 ± 30 years ago, a radiocarbon age that calibrates to the calendar as A.D. 1305–1420 at 95 percent certainty. This would support the notion that the pond was built not by Chinese (nor perhaps even menehunes) but by native Hawaiians well before the time of Columbus. It is probably not a coincidence that *manahune* in the Samoan language means "commoner." In that sense, maybe menehunes did build the pond, but I doubt the one-night part.

So when did Hawaiians invent aquaculture? If it was an idea imported by legendary South Pacific priests, conquerors, or reformers, exemplified in the legends as a fellow named Paao, why are these great ancient ponds primarily a Hawaiian rather than a South Pacific landscape feature? We believe that the fishpond idea is older than Alekoko. The fish fossil record at Makauwahi Cave shows that anadromous fish like the mullet or ʻamaʻama were coming up into the sinkhole pond even before Hawaiian times. Maybe places like this, and perhaps the mysterious saltwater crater pond at Nomilu farther west along this coast, first gave the Kauaians this idea by providing natural examples of huge fish-traps.[3]

In any case, a legendary place even farther west along Kauaʻi's south shore provided a clue to the earlier origin of the mysterious ponds. Kekupua Fishpond at Kapalawai is far less well known to the public than the Menehune Fishpond at Alekoko, our sinkhole at Māhāʻulepū, or the ponds by the Allerton House at Lāwaʻi-kai. The present owners are in their own way quite mysterious—the legendary Robinsons. These folks are the descendants of the Sinclairs, Scotsmen (a woman was their Kauaʻi founder, actually) by way of New Zealand, who settled on Kauaʻi in the 1860s, eventually buying about a fifth of the island and the entire neighboring island of Niʻihau. Among the living Robinsons are Keith, the iconoclastic naturalist and native-plant grower, and his brother Bruce, who runs some of the family businesses. Their cousin Warren Robinson runs others. It was Warren who invited Pila and me to conduct a study of the pond, through archaeologist Hal Hammett, who was preparing a formal report on the area for an ecotourism development proposal.

It was a thrill to be invited onto this very private property to study this obscure little pond, although what we found was a disappointingly small feature overhung with thorny introduced *kiawe* trees. Very little water stood in the basin, and most of it was choked with soft surficial mud, mats of algae, and thickets of aquatic plants. Shaped like an elongate triangle, it measures approximately 790 by 330 feet (240 by 100 m). A causeway of earth and stone separates a small section from the main pond. By probing with a steel rod and digging shallow test pits, Pila and I showed that, to everyone's surprise, this feature is lined in some parts with shaped stones, faced in a similar manner to the other two, more famous aquatic stoneworks on the island, the Menehune Fishpond and a place called Menehune Ditch just inland from Waimea town.

There was not much sediment in this pond, which was really just a sort of impounded spring-hole tenuously connected to the beach via a small ditch. About 6.5 feet (2 m) of muck, clay, and finally sand told the tale, though. About a thousand years ago, it seems, probably a tsunami or some sort of marine overwash or flood scooped out the site. After this, sediments began to accumulate, including fruits and debris from *hala* (*Pandanus tectorius*), a native relative of palms. At a level dated sometime between A.D. 1050 and 1280, the sediments turn to a finer texture, suggesting that the site was impounded then. Thus, we believe this may be one of the oldest man-made ponds in the islands, indicating an origin prior to the arrival of the South Pacific overlords and not so long after the Marquesans, or whoever these mysterious menehune were, first settled the island.

One message was fairly clear from nearly all lowland sites we have studied: microscopic charcoal particles in sediments point almost unequivocally to first evidence for humans on Kaua'i about 1,000 years ago.[4] There is very little charcoal in sediments anywhere on the island before that time, except for an intriguing peak around 4,000 years ago, mentioned earlier, that suggests a prolonged drought. Otherwise, charcoal and rat bones tell the same story: people got here pretty late, later than some of the earlier accounts suggest based on a few dubious radiocarbon dates from the early years before radiocarbon dating protocols were perfected for such purposes.

But the case is not entirely closed. There was one coastal site out of sync. Lida and I took a core of about 20 feet (6 m), dating back around 9,000 years, from a small wetland *makai* (seaward) of the entrance to National Tropical Botanical Garden's Limahuli Gardens, a spectacularly beautiful place on Kaua'i's north shore. Although this low sedge-and-fern-covered site looks like a bog, it is technically a fen, because some water flows through it on its way between taro ponds above and the sea just below. Nestled behind a line of dunes, the site contains all sorts of sediment layers, including probable evidence for several tsunamis or major hurricane overwashes of the site. One occurs about 1,470 radiocarbon years ago, apparently covering the site with gravel and tree trunks. This layer, containing the wood dated to this time, has some tantalizing bits of charcoal. Archaeologists and oral traditions have long held that the very earliest settlements were on the windward north shores of the island, in areas particularly suited to the style of tropical wetland agriculture known to the early Polynesian colonists.[5] Maybe that's what we are seeing here, and maybe not. The tormented horizon of the tsunami deposit is hardly an ideal sediment layer for splitting chronological hairs. So for now, we are sticking with the idea that people arrived about a thousand years ago, which is consistent with what Steve Athens and other archaeologists working on the question of human arrival in the islands are thinking these days.[6]

A lot of the older books on the subject of Hawaiians that address when they might have arrived suggest a time around A.D. 400. This estimate was derived primarily from genealogy, taking the number of generations mentioned in the ancient chants of "who begat whom" and multiplying by an average generational time of 25 or even 30 years. That always seemed too long to some of us, and indeed, estimates using 20 years bring the time of Hawaiian arrival more in line with notions of recent archaeological publications that focus on a time range of A.D. 800–1000 for human arrival in the islands.

EIGHT

A Snails' Tale

THE HAWAIIAN ISLANDS HAVE, among their many claims to fame, some beautiful terrestrial snails found nowhere else on earth. A few areas in the mountains of O'ahu have retained a fair number of them, but many are extinct most other places in the islands. The O'ahu snails are increasingly rare—poster children for conservationists. Kaua'i had some remarkable land snails, including endemic specialties found only on this island. Big showy shells are still found along Kaua'i's beaches, where they have presumably washed down from eroding creek banks. In a few places, notably Ha'ena and Anini Beach on the north shore, Kealia Beach on the east, and Māhā'ulepū and Polihale on the south, one can find these handsome shells of *Carelia, Cyclomastra,* and other snails that are believed to be entirely extinct. I say "believed" because, showy as they are, these snails might remain undetected in some remote place in the interior. But nobody has seen any of the bigger species alive for decades.

Around 2000 we were interested in finding fossil snail deposits, with the help of our good local friend and malacological enthusiast Reg Gage. We gathered shells of extinct snails from crevices in the cliffs at Polihale, Pleistocene red-soil deposits in the Māhā'ulepū Quarry by the cave, and the sand-covered fossil deposits weathering out along creeks at Kealia, Ha'ena, and particularly Anini Beach (figure 16). Anini, that dreamy enclosed lagoon where beautiful wine-red *Carelia* shells often litter the beach, was particularly interesting. Across the road in the high eroding banks of the small creek skirting the polo grounds, my son Alec and I (by this time he was a middle-schooler and an enthusiastic amateur malacologist under Reg's tutelage) found a very rich deposit of *Carelia* and other extinct snails

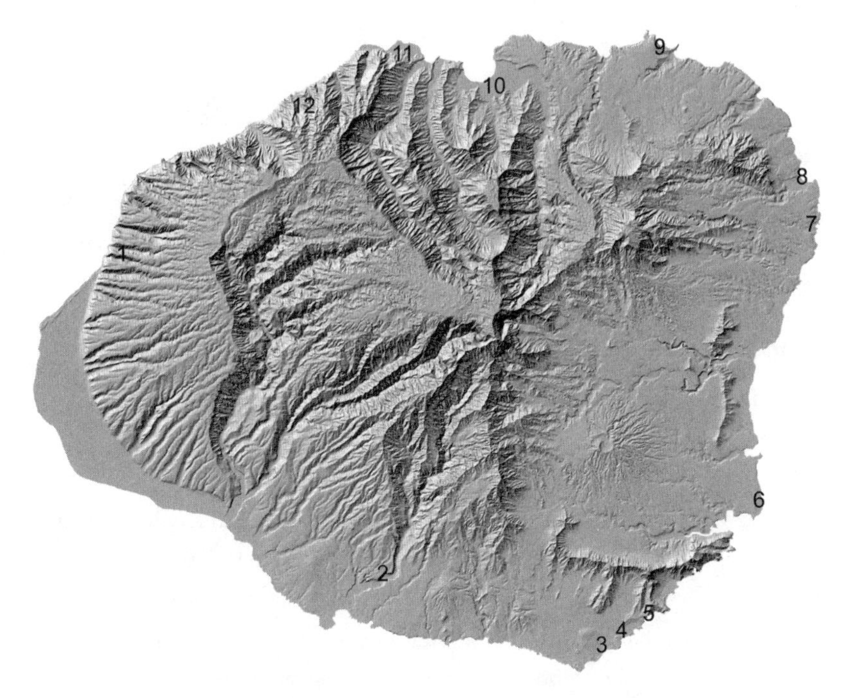

Figure 16. Some land snail fossil sites studied by Reginald Gage and others on Kaua'i: (1) Polihale; (2) Olokele Canyon; (3) Makawehi or Waiopili Dunes; (4) Makauwahi Cave; (5) Aweoweonui; (6) Kalapaki; (7) Kealia; (8) Anahola; (9) Anini Beach and Kilauea; (10) Hanalei; (11) Haena; (12) Kalalau. (Map courtesy of Jonathan Carbone)

Figure 17. The large extinct land snail genus *Carelia* is known only from Kaua'i and Niihau. Many shells are a brilliant wine-red to reddish-brown in color, even 2,000-year-old specimens. This beautiful shell is *Carelia cochlea* from Anini Beach.

(figure 17). The exposed section that Alec and I excavated showed that they seem to have disappeared by the time the first evidence for humans turns up in the deposit, in the form of kukui nuts. A date on one of these snail shells was 1,840 ± 50 radiocarbon years ago. Some of the smaller now-extinct snails persist a little longer, mixing with the kukui nuts. Most likely these snails went extinct shortly after human arrival about a thousand years ago, but it will take more dates to confirm this thinking.

Down the beach at Kealia, there is a creek-bank deposit showing a similar pattern. One medium-sized snail in particular, a species of *Leptachatina,* seems to be mixed in with early kukui nuts, and dates to 1,290 ± 50 years ago. Calibrated to calendar years A.D. 650–870, this one seems like it might come close to overlapping with early humans, depending on the exact date for human arrival one chooses. But these layers were somewhat stirred up by erosion from subsequent marine overwash, and land snails are notorious for dating problems anyway, because they may take up some old limestone-based carbon from the environment in the process of growing their shells. So this information was interesting, but not definitive. Our cave results give a more complete picture, consistent with these scant results from Anini and Kealia, if only from one site.

For a high school science project, our daughter Mara had taken liter-sized samples, at 4-inch (10 cm) vertical intervals, of well-dated sinkhole sediments containing land snails, screened all the shells out, and identified and counted them. This provided a quantitative picture of the changes in the snail fauna. Although disappearance from the sediments is not proof of extinction, it might be a pretty good estimate for extirpation times locally. The pattern was intriguing: larger snails, such as *Carelia, Cyclomastra,* and *Endodonta,* seem to disappear from the record first, more or less at the time of human arrival. Medium-sized snail species go next, about 500 years later as native populations probably reached their pinnacle and some resources, including coastal forests, undoubtedly became increasingly scarce. Finally, the last, smallest species disappeared after the arrival of Europeans.[1] The last native land snail common at the site, a tiny checkered ram's-horn-shaped *Cookeconcha,* disappears in the 1950s, about the time a nonnative carnivorous snail, the rosy wolf snail, appears on the scene, introduced deliberately by the Territorial Department of Agriculture in a failed attempt to control other nonnative snails that had proliferated in the wake of European colonization. We suspect, but haven't yet proved, that endemic birds and perhaps plants also succumbed in "waves of extinction," probably reflecting different causes associated with initial human arrival, subsequent build-up of native populations with resource overexploitation, and finally, a new onslaught of human-derived challenges with the coming of the Europeans in their big ships, officially beginning in 1778 with the arrival of Captain James Cook and those who followed him.

NINE

Mauka Marshes

BY THIS TIME WE WERE BEGINNING to write the script for some parts of the local "ecological theater and evolutionary play," as ecologists since the day of G. Evelyn Hutchinson have been saying.[1] We were piecing the coastal story together, including some key points regarding the human role. But what was the climatic background? Was the late Pleistocene very different, and how has it all changed since then, over the ensuing eleven thousand years or so of the Holocene up to the present? Are the effects of humans on islands like Kaua'i much amplified by climate change, or is that not an important part of the mix? Or is climate, not humans, the "true cause" of the huge biotic changes that occurred around the beginning of the human era, as has been argued for so many other parts of the world with various degrees of success?

To address such questions, we needed to look higher up on the island. The coastal sites didn't seem to show much evidence for big climate changes in the 6,000 years or so that such sites have archived sediments, since the previously rising sea level stabilized near present levels. The island has interior *mauka* wetlands—swamps, marshes, bogs, fens, crater bottoms—that have presumably accumulated sediments for many thousands of years longer.

I had worked in the early 1990s on a 10,000-year record from one of the highest bogs in the islands—Flat Top, at about 7,500 feet (2,270 m) in Haleakala National Park on Maui. This is more than 2,000 feet (610 m) higher than the highest point on Kaua'i. We camped by this bog for three days and flew the core out by helicopter.

Pollen, diatom, and charcoal particle analyses from this site that we prepared (Lida and my graduate students at Fordham, including Bob Decandido, Faith Kostel-Hughes, and Anita Goetz) revealed that the early Holocene was cool and dry. A wetter phase occurred between 5,800 and 2,200 radiocarbon years ago. Since that time, still before human arrival by some centuries, up to the present, the climate high on Maui was moderate and variable. Nothing showed in the record to suggest that climate change has been a bigger factor since human arrival than before.[2]

Over the years there have been a few other fossil pollen studies in the wetlands of the Hawaiian Islands, beginning in the 1930s with the pioneering Swede Olaf Selling. These continue through Jerome Ward and Steve Athens of the International Archaeological Research Institute, Inc. (IARII), and more recently with the excellent palynological work of Sara Hotchkiss of the University of Wisconsin. Each study has shown similar patterns, with only moderate Holocene vegetation change at sites in the full range of elevations on other islands in the chain. As one might expect, earlier millennia, as the last Ice Age waned and the climate warmed to Holocene levels, show cooler climes and some fairly rapid changes.[3]

One place we wanted to core was the highest big wetland on Kaua'i, the vast Alaka'i Swamp of the northwestern interior. Selling had worked there in the late 1930s and obtained some interesting cores that showed approximately the same sort of pattern as our later work on Flat Top Bog, Maui. Despite Selling's wonderful painstaking study of the well-preserved fossil pollen grains, his work preceded radiocarbon, so he could only guess the age, based on comparisons to the postglacial European pollen zones, vegetation trends already well known by this time, and in some cases linked to events of known historical age. As it turns out, Selling had it about right, we found.

Lida and I obtained permission from the state authorities to recore the approximate location of Selling's Kilohana cores. The exact location is unknown, because his work predated the automated advantages of the era of satellite-based Global Positioning Systems (GPS) by many decades. We reached the site by hiking the boardwalk trail across the Alaka'i Swamp to the vicinity of the Kilohana Overlook, where the trail stops abruptly at a breathtaking cliff looking down into the valleys and ridges of the north shore. This was sure to be a great adventure, one of those extreme examples from our long coring experience in which the entire coring rig had to reach the site not by boat, plane, or helicopter, as in many cases, but overland through miles of rough terrain on our backs. We have done this in Kenya, Madagascar—even our native North Carolina—but the logistics were extra-complicated in this case. Our permit stipulated that we could not camp at this fragile site, only work there in the daytime and tramp out each day to our small tent in a designated state-park campground. This project was not well-funded enough to hire a helicopter.

Needless to say, this made for some long days, and heavy packs. Our success at recovering a Livingstone sampler piston core of 10.5 feet (3.2 m) was hard-earned. We worked near a fenced exclosure the state had built to keep pigs from ripping out the native bog plants. Our corer's piston cable was secured on a metal tripod made from three spare fence posts the previous workers had inadvertently left behind for us. We managed to carry everything else in, choose a spot and set up, cover our stuff with a tarp, walk out for the first night's sleep, return at dawn and core all the second day, walk out and sleep, walk back and finish the coring on the third day, break down and pack the gear and coring pipes now full of heavy mud, and carry it all out by nightfall. Not bad for a middle-aged couple! One day a storm blew up suddenly while we were quite busy coring. Lida bundled up, remembering wisely that even in the tropics, getting wet and standing in strong wind can chill you to the point of lethality. I was preoccupied with the work, and kept coring in just shorts and a T-shirt. When vertigo struck, I suddenly realized I was not just tired but very chilled, and felt frighteningly disoriented. As we hunkered down in the shelter of a shrub (it's a vast, mostly treeless wilderness area), Lida found me a sweater and wrapped her arms around me until I stopped shivering and came back to my senses. That was probably one of those times that doing remote field work alone, something I have certainly done more than my share of over the years, would have definitely been a mistake, possibly a fatal one.

Our Alaka'i core showed that the swamp holds at least 16,000 years of ecological history. In the late Pleistocene and up until about 5,000 years ago, clay-rich sediments suggest that the swamp was more seasonal and probably cooler and drier. Permanently wet conditions settled in on this site about 5,000 years ago, with vegetation probably similar to what is there today. There is very little microscopic charcoal in any of these sediments, except the uppermost, probably modern levels.[4]

Putting these results together with Selling's pollen stratigraphy, a pattern emerged similar to that of Flat Top Bog on Maui: after the last Ice Age, it was cool and dry. By mid-Holocene times, it was somewhat wetter, and has continued more or less that way to the present. We later obtained similar results from Wahiawa (Kanaele) Bog, at a much lower elevation, 2,182 feet (665 m), on the flank of Kahili Mountain in south-central Kaua'i, reaching back 23,000 years. At another place called Kilohana, far from Alaka'i Swamp's Kilohana Overlook, we had cored Kilohana Crater way back in 1992 with our friend Don Heacock of the Department of Land and Natural Resources, Aquatic Division. This one, at only 800 feet (243 m) elevation, has a seasonally wet clay substrate in the bottom today that is undoubtedly more like what Alaka'i Swamp must have been like up at Kilohana Overlook in the late Pleistocene and early Holocene, when a comparable heavy clay was laid down. Similarly, pollen was not well preserved in this usually wet, always sticky clay. But below it—wow!—about 5 feet (1.5 m) beneath the gooey surface is wonderful coarse peat, rich with plant detritus and pollen of many of the same plants we detect in our coastal cores. Just below that, the sediments are more clayey, but still fairly organic. A date from just below the changeover to this rich clay was 26,020 ± 230 radiocarbon years ago, well before the last maximum stand of Ice Age glaciers in the Northern Hemisphere. So this dramatic volcano, Kaua'i's biggest crater, is at least that old, perhaps much older, as these rich clays extended down to a very tough hardpan that stopped our coring at about 10 feet (3 m). Like other interior sites, charcoal particles are present almost exclusively in the topmost, essentially modern layers, suggesting that prehistoric fires were extremely rare in the wet interior of the island.[5]

I took a couple of cores from another interesting interior bog, this one a fen site that was surprising for its low elevation. Typical bog plants, including the little carnivorous sundew (*Drosera angliae*) that is found in the Hawaiian Islands and also Alaska (probably the sticky seeds traveled on the muddy feet of a migrating shorebird) grow at lower elevation here—only 555 feet (169 m)—than any other place I have seen in the islands.

To work in this swamp just above the scenic Silver Falls, I needed permission from Ule Ratchener, a wealthy German who owns the property and maintains a trail around the edge of the bog terrain for horseback riders patronizing his commercial ranch. I took a short core there in 1999 with Dr. Adam Asquith, formerly with the U.S. Fish and Wildlife Service, and a longer one in 2000 with the help of Cameron McNeil, a graduate student from the City University of New York who had been studying palynology in my Fordham laboratory. The 11.5-foot (3.5 m) record from this swamp reached back over 8,000 years, and showed changes consistent with the other sites. Notable at this lower-elevation site was some microscopic charcoal in the sediments at times before human arrival. One of these fire-prone periods was roughly 7,000 years ago, and the other we could not date, because it was too stirred up by present-day roots that have penetrated the layer. But the period may correspond to the charcoal peak observed well before humans in various coastal sites about 4,000 years ago. Perhaps these represent droughts lasting several decades that are similar to those documented for other sites around the tropical world at these times, or increased lightning activity, or both.

In subsequent years, we have continued to study and analyze all these sites and work on others that have helped clarify some details. A core from a small wetland on singer-actress Bette Midler's property just inland from Kapaʻa's main shopping center provided a picture, from well-preserved pollen and spores, of eastern lowland forest prior to humans, back to about 7,000 years ago. At Waipa Farmers' Cooperative on the north shore I took cores all around the property, showing that the fishpond there was in fact not old, but rather a historical feature as some elderly locals had claimed. I also saw, from a core far inland behind the barns on a field once cultivated for rice by the Chinese, evidence that the site might have been overwashed by a tsunami about 1,000 years ago.

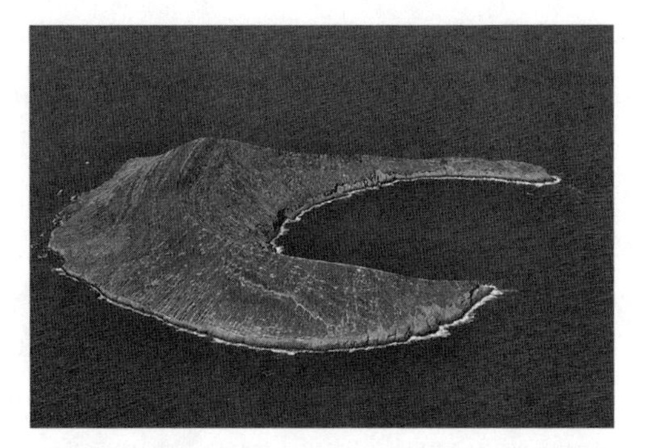

Figure 18. Oblique aerial view of Lehua from the east. This partly flooded crater is a tuff cone formed as the island of Niʻihau began to sink back into the sea floor. At 290 acres (117 ha), this is the largest offshore islet in the main Hawaiian Islands. Although uninhabited by humans, it is home to tens of thousands of pelagic seabirds, including shearwaters, boobies, and albatrosses.

For comparison between modern and prehistoric tsunami deposits, a big stranded sea cave on Kauaʻi's north shore at Haʻena proved to be an interesting place to see the results of a big modern tsunami in a cave setting. On April 1, 1946, a tsunami over 50 feet (15 m) high slammed into Haʻena, killing several people nearby and dumping truckloads of sand, gravel, cobbles, and even large stones into the front of the Haʻena Dry Cave.[6] Two other stranded sea caves nearby contain interesting freshwater lakes. The lake inside one of the wet caves was 52 feet (16 m) deep, we discovered to our surprise.

There are many other interesting places on this island where we still haven't looked yet, and should, to flesh out the geographic extent of the voyage of the Poor Man's Time Machine. Nevertheless, we like to think that the proverbial blind men have felt the elephant in quite a few ways by now.

But we still haven't really nailed down the fungus story. *Sporormiella* has scarcely appeared in the Kauaʻi sediments, before or after the arrival of humans. Mostly we have seen them there in modern sediments near cow pastures, which is no big surprise. But we have seen lots of them in the sediments from a small cave on Lehua Islet, a roughly 1-square-kilometer rock off Ni‘ihau, Kaua‘i's neighboring island (figure 18). I accompanied federal and state officials on a research trip there in July 2004 to excavate an owl-roost deposit. This tiny island is literally elbow-to-elbow with nesting seabirds such as boobies, tropicbirds, shearwaters, albatrosses, and frigatebirds. A succession of generations of the European-introduced barn owl had made a thick pile of owl pellets and dung, including thousands of rat and rabbit bones, in a small cave. These "sediments" contained vast quantities of *Sporormiella*, and little else in the way of microfossils. Wood from a prehistoric hearth and pollen from a long-buried spring seep on the island revealed that the vegetation was a lot richer there before rabbits introduced in the past century ate it up. And a date on the wood excavated by state archaeologist Alan Carpenter and his colleague Tom Dye showed that people had built a fire there, perhaps on a trip to collect birds and bird eggs, about eight centuries ago, when early Polynesians were probably still just settling in on Kaua‘i and adjacent Ni‘ihau.

TEN

So What Happened, Anyway?

WHENEVER THEY ARRIVED, whether a millennium ago or a few centuries more than that, and whoever they were, the first people to see the island of Kaua'i saw a world that has almost completely disappeared. Leeward lowlands that are now entirely without native forest hosted a diverse assemblage of trees comparable to the most pristine interior highlands. Except for bats, the only terrestrial vertebrates were birds, some quite large and many now extinct, like the turtle-jawed moa-nalo (*Chelychelynechen quassus*), a flightless duck. After all this research in more than a dozen sites around the island, we can still only guess at the workings of ecosystems in which bird-catching owls (*Grallistrix auceps*) were the top carnivores, large flightless geese and ducks were the grazers and browsers, rats and ants were absent, and land snails of many types, mostly unique to this island, were ubiquitous.

That world is almost completely gone, represented by some surviving plants, invertebrates, and birds, but missing many important characters. Our radiocarbon dating of bones of extinct birds, shells of extinct land snails, and the sedimentary layers containing these lost souls shows that many if not all survived into the human period, but not long after. Although the record shows clearly that some native species persisted until European times, then dropped out, it also shows that the coming of the Polynesians, despite their traditional reverence for Nature, was a catastrophe for the long-standing endemic biota of the islands.[1]

Catastrophe, that powerfully negative word. Some dear Hawaiian friends will wince to read that word as a description for what happened after they reached the islands, but no offense is intended. What happened is really no different from what transpired everywhere that humans have reached in the past 50,000 years as they spread from their African and Asian homelands. Why are newly arriving humans so devastating for environments that evolved in the absence of humans?

The arrival of the third millennium was a timely opportunity to be thinking along these catastrophic lines. A quick look back over the past decade, beginning with the "Y2K" fears that peaked on New Year's Eve of 1999, serves to remind us that humanity has become somewhat obsessed with the whole idea of disasters. Although it is fashionable to blame the news media for this fearful negative mentality, some awful events have indeed been major historical turning points, starting with the September 11, 2001, attack on the World Trade Center in New York and the Pentagon in Washington, D.C. This date was, ironically, the ninth anniversary of Kaua'i's biggest modern catastrophe, Hurricane 'Iniki. I was at home in New York then and, like most people, will never forget what I was doing when the news came. I am almost certainly the only person you will ever meet who can say that, when 9/11 happened, he was picking mushrooms. In our backyard in the tiny village of Croton Falls, in Westchester County north of New York City, Lida and I had a big stack of spore-inoculated oak logs on which we had been cultivating for some years successive bumper crops of delicious shiitake mushrooms (*Lentinus edodes*), a gourmet's delight. On that lovely warm September day, the nearby windows were open, and I abruptly stopped my pleasant work when Lida shouted something about a plane crashing into the World Trade Center. I went in to turn on the TV just in time to see that horrible sight that is burned into the memory of nearly every human in the western world and beyond—the second plane plowing into the remaining tower.

For days after, as the pall of smoke rose from the ruined towers, distantly visible from windows at the end of my floor in Fordham's Larkin Hall in the Bronx, bad news and a mixture of hurt and anger swirled up for New Yorkers and the world. My son's Little League coach was a fireman who died heroically in the rescue effort, a neighbor's fiancé was unfortunately at work on time in his office when one of the planes came right through the windows of his floor, and the husband of a former graduate student in our department was among those who didn't make it out. It was a time of sadness and confusion for the neighborhood and the entire world, and history turned a sharp corner that September.

The summer before, we had completed a film at the cave and all around Kauaʻi with the WGBH-Boston group that creates the television program *Nova*. We were in the third episode of a seven-part series titled *Evolution: A Journey into Where We've Been and Where We're Going.* Our part was about extinction, not just as a natural process that moves evolution along by clearing the way for new species (like the dinosaurs giving way to the mammals at the end of the Cretaceous), but also as a catastrophic truncation of many evolutionary lines with the coming of humans to a pristine landscape. The television debut of this series was yet another casualty of the 9/11 mentality, because the public's attention was far from programs about natural history, no matter how well prepared. Of the many films Lida and I had worked on over the years, we felt this was the best to date, but it went largely unnoticed until it was rereleased the following year.

In the meantime, I had been helping the British firm Darlow Smithson to put together another film on this topic—a grandiose two-hour production filmed all over the world—called ultimately *What Killed the Mega-Beasts?* I had been working on the story board and script for this film off and on for months, but was a little surprised that the firm wanted to move ahead with filming my half-hour bit in Madagascar just a couple of weeks after 9/11, not long after planes began flying again. Friends and family wished me good-bye and good luck with more fervor than usual, it seemed, perhaps because I was one of the first people they knew to venture out into this not-so-brave new post-9/11 world, and I was going to a remote, always vaguely dangerous place virtually next door to some of the trouble spots of the Arab world. So I flew over on an eerily empty Air France jumbo jet to arrive in a Madagascar awash with strange local rumors that Osama bin Laden was thought to be hiding out in their island country, and that the Americans might invade them any day . . .

The filming was a pleasant distraction from these dark times. Once we got out into Madagascar's remote spaces, the world's turmoil seemed like something on another planet. We settled into filming my Malagasy colleagues and me not just digging and finding fossils, but doing some pretty exciting stunts, like rappelling into deep dark pits in a remote cave. We sailed a huge creaky old wooden boat of a very traditional style that we think might be similar to the ones that carried the first Malagasy to the island two millennia ago. We also filmed a huge crackling rain-forest fire from a small plane just above, in the smoke, with the door off and the cameraman strapped to the struts so he didn't fall out. It was great fun.

A few weeks later I returned to New York. The catastrophic events of recent weeks were still very much on everybody's mind, as the World Trade Center cleanup continued and the lists of the dead grew. Fordham University alone had more than thirty staff members, students, and alumni who died in the conflagration, and rumors of long-term environmental effects on surviving New Yorkers were rampant. By this time, the hunt for Osama bin Laden was in full swing in Afghanistan, and talk of expanding the war to Iraq or perhaps other places was all over the media.

A year of public fears of impending additional cataclysms followed. By the next September, I was at a conference at Brunel University in London that was devoted entirely to the now trendy topic of catastrophes. Organized by some of my European colleagues, this excellent meeting was a turning point in my own thinking. All kinds of scientists and professionals who deal with catastrophes (from astrophysicists who model the likelihood that the earth will be struck by a big extraterrestrial object, to geologists who study volcanic explosions, earthquakes, and tsunamis) delivered papers. There was even a fascinating talk by an actuarial statistician who had been calculating the odds that something else really bad would happen soon, and what the costs would be, for insurance companies. I was one of the few ecologists invited, essentially to give one of my lectures on "humans as an ecological catastrophe" using examples from my research around the world on this topic that had suddenly become somewhat interesting in this new light (or darkness, perhaps) of a world obsessed with disaster. Of course, many of the speakers reminded participants and the media covering this that catastrophes of many sorts happen all the time, and despite their inevitability, we are never in a sense really ready.

That depressing thought has followed me through the subsequent years. In the ecological realm, for instance, it seems that human arrival is inevitably, as Ross MacPhee of the American Museum of Natural History describes it, a "deadly syncopation."[2] People come, and ecosystems collapse. Why is this outcome inevitable? What is it about being human that is such a disaster for nature every place we go? More than ever, I wanted to know. But in the dark days of the new millennium, I was increasingly challenged by another notion potentially more optimistic: if we could really understand what has happened in the past, could humans do better by their ecosystems in the future? Or is this "deadly syncopation" the only possible outcome, and a trend that humans will extend to every corner of the planet, and all other at-risk species, as the ensuing decades unfold?

The litany of negative things people do to nature is not that long, really. Scientists studying extinctions around the world have identified three main categories: overharvesting, landscape transformation, and biological invasion. The first is the most obvious, perhaps. Stories abound of people hunting a species to death: the great auk, Steller's sea cow, and perhaps that first of all historically documented extinctions, the dodo.[3] But I looked into the dodo a bit while visiting some ecological restorations on its native Mauritius a few years ago, and came to realize that even this seemingly clear-cut case is more complex. It turns out that the Dutch and other early Europeans on this previously uninhabited island soon discovered to their disappointment that this naive giant flightless pigeon was not that tasty, and they lost interest before they finished the job. But stray dogs, feral pigs, and introduced macaque monkeys apparently did in the rest. Rats could have played a role, too.

Any hunter can tell you that, although some members of a species may be easy quarry, there are nearly always some that are far more wily—especially ones that have been shot at and missed, or have escaped from a trap. So ideas about overhunting as a route to wholesale prehistoric extinction, as with three-quarters of all large animal species in North America, generally turn on the notion that the process has to happen fast, before creatures realize what is happening and begin to fear humans or any other introduced predator and add avoidance to their behavioral repertoire. That is Paul Martin's Blitzkrieg Hypothesis in a nutshell.[4] This is easier to visualize on a small island, partly because of the smaller land area and animal populations, and partly because that is the primary place where historians have recorded the phenomenon—Mauritius and other Mascarene Islands, the Galapagos, and a handful of other remote islands being cases in point. But the phenomenon also still exists on a bigger scale, too. Penguins and seals in Antarctica, and whales in some remote parts of the open oceans continue to be more trusting of humans than is good for them, for instance. I have seen lemurs in really remote forests of Madagascar, when approached by a human, come *down* the tree, rather than moving up—potentially a lethal mistake.

It's not hard to imagine that the big flightless ducks and geese of the Hawaiian Islands responded in such a foolish way to the first Polynesians. As adults, these turkey- or swan-sized birds probably had no plausible natural predators. And humans, after all, don't necessarily *look* all that dangerous. Tall, slender, and smooth-skinned, they must have looked like another flightless and ludicrously featherless bird to a moa-nalo (Hawaiian for "lost fowl," a term cooked up by Storrs Olson, since there is no word for these long-dead ducks surviving in the Hawaiian language). It's not hard to imagine that any of the larger birds of the extinct fauna, on their first and last encounter with a hungry Polynesian, failed to flee and turned into a kind of instant luau. That's not to say that these people were in any way bad or wasteful, they were just feeding their families like anybody else, and had no way of knowing that the cumulative effect of their actions would spell doom for a species (remember, right up until the time of Thomas Jefferson, even most educated Euro-Americans still denied the possibility of extinction, since God wouldn't have allowed for such imperfections in creation).

So hunting and otherwise overharvesting (egg and plant gathering, for instance) might work for some of the larger birds on Kaua'i, perhaps even a rare and geographically restricted plant or invertebrate. But all kinds of

Plate 1. Down in a hole on the northwest side of the Makauwahi Sinkhole, David Burney excavates extinct bird fossils from the mud without tools, using pumps that allow him to dig far below the water table in reasonable comfort. (Photo by Alec Burney)

Plate 2. (left) Lida Pigott Burney harvests the brilliant orange pods of the white-flowered maiapilo (*Capparis sandwichiana*). This rare and beautiful native plant was common in the record of fossil pollen and seeds at the cave. Besides the one plant that survived in the sinkhole, the Burneys and hundreds of volunteers have planted many more of this and approximately 100 other native species in the Makauwahi Cave Reserve. (Photo by Alec Burney)

Plate 3. (above) Storrs Olson and Mara Burney excavate a fossil bird from an exposed bank in the Makawehi Dunes near Makauwahi Cave. (Photo by Lida Pigott Burney)

Plate 4. Makauwahi Sinkhole as it might have appeared from the southwest 3,000 years ago, before humans arrived, in a painting by Julian Pender Hume. At left, a flock of turtle-jawed moa-nalo graze kawelu grass and the orange fruits of maiapilo. Palila and a flightless rail are on the ground in front. Immediately above them is a Kaua'i o'o. A flock of the medium Kaua'i goose with fighting males is in the background.

At right, the extinct long-legged Kaua'i owl scatters Kona finches. Landing on the pond are Laysan teal. Flying overhead, from left to right, are i'o or Hawaiian hawk, frigatebird, Newell's shearwater, nene, and Hawaiian petrel. At lower right, on the leaves of *Kokia kauaensis,* are *Blackburnia* beetle and *Rhyncogonus* weevil; on the trunk of *Pritchardia* palm, a hoopoe-billed akialoa and endemic *Carelia* land snail.

Plate 5. Artist's reconstruction of an extinct native finch (upper) discovered at Makauwahi Cave and described by Storrs Olson and Helen James. They named it *Loxioides kikuchi* in honor of Pila Kikuchi. The common tongue-twisting name that has been attached to this bird is "Pila's palila." Also pictured (lower) is the conventional palila (*L. bailleui*), whose bones have been found at Makauwahi Cave, too. (Painting by John Anderton, copyright Smithsonian Institution)

Plate 6. Looking south from the Makauwahi Sinkhole rim in 1996. This was nearly four years after the collapse of a huge banyan inside that was thrown down by Hurricane ʻIniki. A massive tangle of invasive nonnative vegetation took over from the toppled giant, completely obscuring the view of the limestone features around the rim.

Plate 7. Same view 12 years later, in 2008. Since planting in 2002, the palms have shown remarkable growth rates, now poking their crowns out the top of the sinkhole. A wide variety of native trees, shrubs, and groundcovers has been established. Most of the plants chosen are prominent in the site's fossil record.

smaller and less edible creatures went extinct, too, and the pollen records here and throughout the world have shown just how profoundly humans can change environments with fire, axes, and cultivating tools.

The journals of Captain Cook and other early European visitors are replete with references to the extent that Hawaiians had transformed the land before Europeans had a chance to do so. When Cook came ashore at Waimea on Kaua'i's west side, he and some other crewmen took a long stroll inland. They were amazed to find that every place they looked, they saw tidy agriculture—taro pondfields, and orchards of banana and breadfruit. When one of his lieutenants, George Vancouver, returned a few years later as captain of his own ship, he was quite frightened at night to see the light of so many campfires twinkling all over the hills above Māhā'ulepū. Although these were undoubtedly just people's hearths, he instead opted for the paranoid interpretation that the Hawaiians were gathering from all over the island to attack his crew.[5]

Anthropologists and historians have repeatedly asserted, and provided plenty of indirect evidence, that when the first Europeans arrived, the population of Kaua'i was probably as high as it is today, perhaps much higher.[6] No matter how good they might have been as stewards of the land, so many people living off the land were bound to have had an effect. Conventional wisdom, derived from oral traditions, modern plant distributions, and now also the fossil pollen evidence, suggests that Hawaiians before European contact had mostly long since converted the lowlands to their purposes. They probably left the higher, wetter, and cooler mountains more intact, using these more remote areas primarily for bird hunting, limited tree harvesting, and quarrying stone for their tools. Our own results from paleoecological sites all over the island, as well as that of others throughout the islands, support these notions. In a number of coastal sites, I have also been able to demonstrate that erosion rates increased somewhat in coastal sites after the build-up of Polynesian populations—but nothing like the erosion rates of the plantation era of the past two centuries.[7] Our results from fossil seeds at Makauwahi Cave also confirm that, even centuries after Polynesian occupation of the landscape, many, but not all, native tree species that were there before their coming still persisted. So it's easy to imagine that prehistoric landscape transformation, as well as overharvesting, might have played a role in the big changes of the past millennium on Kaua'i, but there still must be more to this recipe for disaster, we suspect.

The other big factor, human-assisted biological invasions, probably goes a long way toward explaining the rest of the pattern. Hawaiians brought pigs, dogs, chickens, and the Pacific rat, as well as smaller stowaways like gecko lizards and various insects and snails. These each could have had profound effects: pigs and dogs are well-documented predators of ground-nesting birds (remember the dodo); chickens could have been tough on native invertebrates and perhaps competed with some native birds; and even the small, cute Pacific rat has been observed to have devastating effects, by eating seeds of native plants as well as invertebrates and bird eggs and young. This little rat continues to be a threat to nesting seabirds on Lehua Islet off Ni'ihau, for instance. Its days are presumably numbered, however, because rat eradication is slated for the near future, now that the federal and state biologists are finished exterminating the rabbits.

Some devastating biological invasions caused by humans on Kaua'i are essentially invisible, yet no less profoundly damaging. In 1826, whalers introduced the mosquito *Culex quinquefasciatus* to the Hawaiian Islands, probably by dumping their shipboard water casks and refilling them in the local streams. These mosquitoes are annoying enough (when I swat one I sometimes reflect on how much more of a paradise the place must have been without them), but they also carry microscopic menaces such as avian malaria and avian pox. It is not certain when pox was introduced, but avian malaria is thought to have come into the islands just before World War II, perhaps with introduced birds. These diseases are fatal to native landbirds, including the honeycreepers and other species so numerous in our prehuman and Polynesian-aged cave deposits. As a result, today on Kaua'i there are no honeycreepers or other native landbirds below about 3,300 feet (1,000 meters), the approximate elevation limit of the mosquito and the diseases that hitchhike with them. That limit seems to be rising in recent years, perhaps in response to global warming. Many native landbirds whose bones are common in the Makauwahi Cave sediments, including the oʻo, oʻu, and akialoa, are probably now extinct on Kauaʻi, at least in part owing to this and other disease challenges.[8]

So do these three classes of gremlins—overharvesting, landscape transformation, and biological invasions—add up to enough bad stuff to account for Kauaʻi's biotic collapse? Probably. But places far larger than Kauaʻi, from Hawaii's own Big Island to Madagascar to the Americas, all show the same catastrophic pattern, and likewise pretty soon after the first evidence for humans, and running far more deeply into ecosystems than one might have predicted from these causes alone. In that respect, modern Kauaʻi sheds some light on all these places, because here, sadly, the extinctions are still going on. I witnessed one in 2006. The last known individual of a beautiful native lobelioid plant, *Cyanea kuhihewa,* died in the native plant greenhouse at the National Tropical Botanical Garden in the Lawai Valley on Kauaʻi's south shore, a few steps from my office. Like other historical extinctions, and presumably prehistoric ones as well, more than one problem did it in. When NTBG botanists first discovered it a few years ago in Kauaʻi's remote interior, the poor plant was already rare, probably because of habitat reduction and degradation by humans, and the environmental damage of feral pigs. Introduced slugs and nonnative insect pests were probably also chomping away, and it was being crowded out by aggressive introduced weed plants in its habitat, such as Koster's curse (*Clidemia hirta*). Hurricane ʻIniki, the one problem that humans (presumably) had no part in, devastated its habitat and left the area open to further invasion by weedy competitors. And a mysterious disease, probably a root-rot fungus, killed the last plant. So what actually caused the extinction of *Cyanea kuhihewa*? The answer is probably all or several of the above, and perhaps other unrecognized agencies, all interacting to reduce population sizes to zero. The same is probably true for many other recently extinct species here and around the world—and by extension, whole ecosystems and biotas for thousands of years. This is the phenomenon of *negative synergy.*

As a veteran of many debates regarding extinction causation, in the literature, at scientific meetings, and on the lecture circuit, I have had to take some flak for this idea. My friend, sometimes ally, and sometimes adversary in scientific debate, Ross MacPhee (who was also my postdoctoral adviser), calls this synergy idea "the intellectual low road." He and many others in the debate, understandably, don't like these combinatorial solutions, preferring to argue "Cause A versus Cause B versus Cause C." While I agree that "all or some of the above" is not a philosophically elegant answer, I think it has some usefulness. First of all, systems theorists, and plenty of everyday people, tell us that almost nothing in the real world actually happens for a single reason. Whether it is combating AIDS, explaining how Carolina Bays and other oriented lakes form, or accounting for the evolution of language, organisms, or stars, it is pretty certain that more than one thing has to happen in concert to get the desired effect.

This idea of the negative side to synergy as an explanation for extinctions was elegantly codified for me by an off-duty commercial jet pilot who was sitting next to me on a flight some years ago. We got onto the subject of what we each did for a living, for me a somewhat more difficult question to answer perhaps than for an airline pilot. Eventually I got to the point of explaining my lifelong interest in extinctions, and my theories derived from examining events around the world and through the millennia. I told him that I believe that most of the big catastrophic extinction events of human time are caused not by a single primary human impact but rather by the interaction between several human impacts with natural climatic and ecological dynamics. Seldom, I opined, does just one thing go wrong, and when just one has, the effect is probably much less powerful and may even go unnoticed.

"That's exactly what the experts have concluded from studying airplane crashes," he replied. His explanation went something like this: With the exception of terrorist bombings, very few jetliners ever go down for a single reason. These big planes fail mechanically all the time, and passengers don't even know it. Crashes occur when, simultaneously, more than one thing goes wrong. For instance, a mechanical failure occurs during a storm, or a system fails and then the backup fails, too. Or a mechanical failure is made worse by a pilot error, or vice versa. Ultimately, planes hit the ground simply because they fall out of the air, and it usually takes a combination of problems to bring one that low.

I can't think of a better analogy than that to explain how humans can be such an ecological catastrophe. Humans cause wholesale extinction and transform ecosystems almost overnight not because of one thing they do, but because many of the things they do make other things worse.

Is this a useful notion, or too obvious and fuzzy to help, as Ross suggests? Here are some other questions I have wanted to ask that arise from this one:

How inevitable is the negative synergy of human-caused extinction?

If people are the problem, can an awareness of this lead them to also be a solution?

These are not just questions that I have an academic interest in answering. These are fundamental to all that is done in the name of conservation. As the years have ground on at the sinkhole, a series of events, some good, some not so good but inevitable with the passage of time, have made these questions my underlying theme, and the motivation for much in the subsequent history of that place. What started for me as dark musing about catastrophes at a time when the whole world was doing the same has turned into something of a personal crusade.

ELEVEN

Greetings from Old Kaua'i

AFTER A WHOLE CAREER OF THINKING about the distant past, it might seem strange that I found myself, in the early twenty-first century, thinking mostly about more recent times, even the present and the future. Global events certainly had something to do with this, but local developments on Kaua'i played a very big role. As my colleagues and I uncovered and integrated all this new information about Kaua'i's more remote past, I had become acutely aware that one of our greatest temporal blind spots down at the cave was, in fact, the more recent past. There were a lot of mysteries still to be unearthed not just in the sediments of the sinkhole and our other sites, but in old books, maps, and photos. Most important of all, in hindsight, there was much to be learned from talking to those *kupuna* (elders) who had been in the vicinity their whole lives, absorbing old stories and remembering their own experiences from "old Kaua'i."

I was beginning to feel a certain urgency in this pursuit of the unwritten histories in the memories of those around me, because these treasure troves were slipping away. Adena Gillin, Grove Farm engineer Elbert Gillin's wife and our neighbor in the one house near the cave back in the early days of our work there, told us many interesting stories about the area before she died, but there were undoubtedly many more we had missed. Folklorist Gabriel I (pronounced "ee") had related charming tales about the Māhā'ulepū area, but he was gone now.

And Pila was dying. Just about the time he was set to retire from his lifelong post at Kaua'i Community College, he was diagnosed with prostate cancer. He underwent surgery with his characteristic good humor, joking about being "held together with staples" and so forth. But some months later, he was diagnosed with bone cancer, and underwent long hard bouts of chemotherapy. Despite hair and weight loss, and more pain and discomfort than he would ever publicly admit, he kept on working on our project, which he saw as his last and best contribution to science. In his last few weeks, during the summer of 2003, he and his wife Dolly completed our voluminous report to the Hawaii Department of Land and Natural Resources, Historic Preservation Division, on the archaeological and ethnographic work in connection with the cave project. This was chockful of information regarding the oral and written history of the site that we had been collecting for several years, and filled a lot of gaps in a paper I had been writing with him on the entire human story at the cave.[1]

Pila died at home, with his family around him, while I was on an expedition to a very remote part of Madagascar. Back in Fort Dauphin for supplies, I got the news by email, along with a request to provide a short eulogy to be read at the memorial service. As things like that go, it was quite easy to write, really, despite the sad news. Pila was the kind of person that generated a lot of fond remembrances for everyone around him. His influence on our project, and on a host of people on Kaua'i and throughout the islands, was immense and positive. That's basically what I had to say, and state parks archaeologist Martha Yent, a frequent volunteer down at the cave, read my e-eulogy at the funeral.

We were thankful that not all our sources of oral history disappeared with Pila's passing, but we knew time was getting short. LaFrance Kapaka-Arboleda continued to supply precious family stories that go all the way back to her great-great-great grandfather, and her own rich childhood memories of this place. She frequently visited the cave and shared her stories with us right up until shortly before she died of thyroid cancer in May 2005.

Just when we think we've heard and seen it all, another local person shows up down at the cave for one of our tours with a new story, or an old picture from those remote times "before 'Iniki." I like to joke that, whereas most of the world measures time as A.D. or B.C., here on Kaua'i everything is either before 'Iniki or after.

Before launching into the oldest stories from the cave, I need to clear something up. Throughout this book I have used from time to time the term *prehistoric* for those times before the beginning of written history on Kaua'i, which officially starts with the journals of Captain Cook and his crew. Some of my Hawaiian friends are a little sensitive about this word being used to describe the times of their ancestors here. They point out that the Hawaiian people had very detailed oral histories that reach back to their earliest times in the islands and perhaps even before they made the great journey from the South Pacific. Much of this history was written down soon after the advent of writing in the islands, by scholars of these traditions such as David Malo and David Kalakaua.[2] Perhaps the greatest limitation of these histories, as pointed out earlier, is that the chronologies are based not on years from some specific point such as the birth of Christ (or Mohammed, or Hurricane 'Iniki) but rather on generations. This has led to possible cumulative error, as one gets back to earlier and earlier generations, since conversion to years requires mostly unsupported assumptions regarding the length of each generational interval. Newer ideas, tempered by the radiocarbon results and ethnographic observations, point toward shorter generation times, so that some events in oral histories may have occurred later than projected in former estimates.[3] In that sense alone, some of the stories I relate below are "prehistoric" in that we don't know exactly when they happened, only that they may reach back a long time before the late eighteenth century, since Hawaiian families are quite adept at preserving the details of their own histories and genealogies.

There are legends associated with the Māhā'ulepū area that are said to reach back perhaps as far as the fourteenth century A.D. These earliest stories from the cave area suggest substantial human populations along this coast many centuries ago, because they involve inter-island warfare, in which Kaua'i is victorious, and a resource conflict, in which a chief's fishing rights are usurped and poachers are punished.

The first of these stories, traditionally thought to have occurred in the 1300s, concerns a battle that began at Māhā'ulepū, between the forces of Kaua'i's ruling chief, Kūkona, and Kalaunuiohua of Hawaii, who had already conquered the chiefs of Maui, O'ahu, and Moloka'i. Kūkona defeated

Kalaunuiohua by continually retreating into the interior until the invading forces were quite weary and spread out. Then his forces ambushed them in a steep canyon.

The second story, said to be of great antiquity, tells of a giant crab in the caves at Māhā'ulepū and implies a conflict over scarce food resources. It was recorded in 1885 by a student at the Lahainaluna School on Maui interviewing an elderly relative from Kaua'i, as part of a class assignment, and is archived with volumes of these "Lahainaluna Papers" in the Bishop Museum in Honolulu.

> Waiopili was another food altar. It was located at the source of the stream of Mahaulepu. It was an altar for the purpose of multiplying food plants, dedicated to the god Kanepuaa.
>
> Kailiili was a fish altar that stood on the beach at Paa. Near this spot is a long cave with a hole just above it. A famous legend is told in the olden days of the big crab, Kiakala, who belonged to the land of Paa. He took the octopus of the sea at Paa and carried them secretly to the shore for his keepers who lived in the cave. The octopus was prohibited for the use of the chief of that place, who was named Keakianiho.
>
> The crab was a red one. He seized the chief's octopus to feed his people with. Kaneakalau was the kahuna at that time. The chief was puzzled at the lack of octopus when it was time to go spearing them.
>
> When Kaneakalau ascended till he came to the place where he could look down into Kipu, he rested on a mound and fell into a trance. He saw a crab take the chief's octopus and give them to his people. When he came to, he saw that he had been in a trance. He went to tell the chief and when they went to investigate, they found the cave and the crab's people. They were all killed but the crab was never caught.
>
> This was one of the famous legends of this district of Koloa in the olden days.

The Waiopili Heiau mentioned here is located in the limestone quarry adjacent to the cave. This unusual stone temple was described in an early archaeological survey as having walls constructed in an atypical style.[4] Much of it was dismantled in the course of the quarry operation, and some of the materials are said by locals to make up the line of whitewashed stones that adorn the edge of the road adjacent to the Hyatt Hotel nearby.

The red crab is a fascinating detail. One of the very common fossils in the cave sediments, up through early Polynesian times, is a now extinct land crab. Its well-preserved shells and claws often still retain a reddish-brown color, and a closely related rusty-red species (*Geograpsus geayi*) still survives in the South Pacific and all the way across the Indian Ocean to Mauritius. It is interesting that the "crab people" who were poaching the chief's octopus were caught and killed, but the crab was never caught.

My son Alec did a middle school science project involving these crabs. With digital calipers, Alec painstakingly measured hundreds of claws of this extinct crab from layers of the cave's sediment spanning several millennia. What he found was that the average size of the claws began decreasing after human arrival, reaching their minimum size just before the crabs went extinct a few centuries ago. This trend, diminishing average size in an over-harvested species before extinction, is a pattern that has been documented in many places for extinct and endangered species that grow throughout life, from extinct Australian marsupials to tusk size in African elephants, to depleted commercial fish stocks.[5]

So maybe this old story, and Alec's results, provide evidence that, well before European times, human populations in the islands were not only large, but resources were so scarce that some native food items were eaten to oblivion. These crabs must have been tasty—Storrs Olson tells about gathering and eating delicious land crabs easily caught on remote Atlantic islands in his younger days. Alec also looked at changing shell size in other Hawaiian invertebrate food items. The shells of the tasty 'opihi (*Cellana exarata*), for instance, a kind of limpet, start out large in early Polynesian times, decrease in average size until about Cook's arrival, and then start to increase again in the nineteenth century, probably because native Hawaiian populations crashed to small numbers after Eurasian diseases were introduced. This had the effect of giving the 'opihi a break from overexploitation, since few foreigners know how good these mollusks are.

The upper levels of sediment in the sinkhole's East Pit and Middle Pit told us a lot about Polynesian lifestyles and diet, but the richest site for human artifacts was the large pit we excavated just inside the mouth of the South Cave. Pila and the KCC Anthropology Club had the responsibility of starting this pit in 1997. Once a wide hole had been opened down to the water table, however, I moved the pumps over there and dug down over two meters deeper, and far back up under the west wall of the cave, where small underwater grottoes had collected some amazing things. Together, by 2006 the digs in the cave had yielded 89 fishhooks made of pearl shell or bone, and a single one of iron, as well as other fishing gear, bone picks, coral and urchin-spine abraders, stone tools, ornaments, cordage, gourds, and worked wood (figure 19). The most unusual part was the abundance of perishable artifacts, including pieces of canoes and canoe paddles, tattoo needles, adze handles, spears and daggers, and pieces of rope and string from natural fibers, in various braid patterns and often still containing centuries-old knots.

The sinkhole lake was essentially a huge midden during Polynesian times, in which food refuse accumulated below the water surface in oxygen-free mud—an ideal opportunity for us to see what was on the menu in the vicinity for centuries. In sediment layers from the human period we have found remains of many species probably used as food items, including marine and terrestrial invertebrate shells, bones of marine fish and introduced mammals and birds, and the seeds, fruit capsules, and pollen of Polynesian-introduced plants.[6] My favorite food find was the pickled remains of a bitter yam (*Dioscorea bulbifera*), a Polynesian "famine food" gathered from dry sandy areas (figure 20).

Some of the archaeological finds in the cave were quite mysterious. We soon began to realize that, as one might expect for such a dramatic setting, Hawaiians were using the place for some kind of ritual activity. In the first place, there were some remarkable ornamental artifacts. Digging

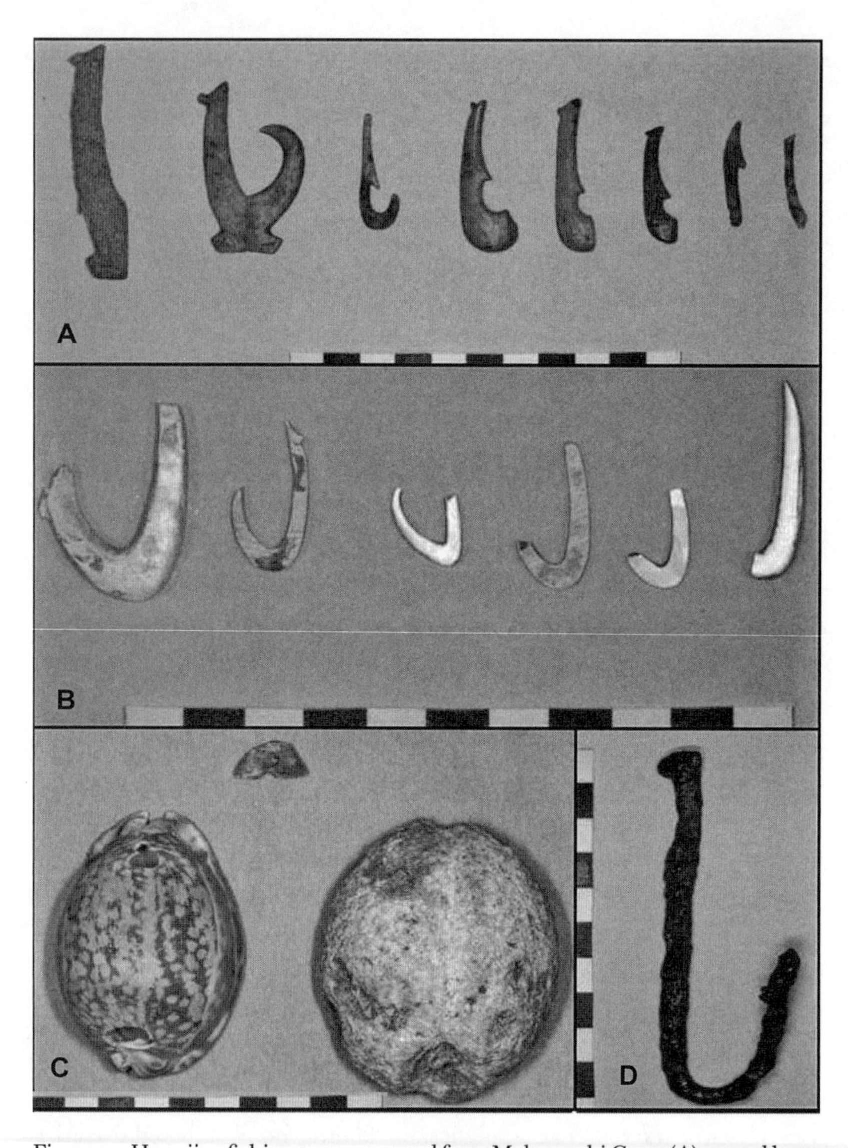

Figure 19. Hawaiian fishing gear excavated from Makauwahi Cave: (A) carved bone fishhooks; (B) fishhooks carved from pearl shells; (C) octopus lure parts, including a drilled cowrie shell, a carved stone toggle (top), and a grooved stone weight; (D) hammered iron fishhook, an artifact from after the arrival of Europeans.

Figure 20. A unique centuries-old "mummified" yam (*Dioscorea bulbifera*) from the Polynesian-era mud of Makauwahi Cave. This is a plant thought to have come to Hawaii with the earliest Polynesian settlers.

under the South Cave wall, I found a really rare and precious artifact—a perfectly round, highly polished prehistoric stone mirror. Yes, *a mirror.* Before European contact, Hawaiians made mirrors by painstakingly polishing a piece of dark, fine-grained basalt to make a perfectly flat, glass-smooth surface. Holding such an object horizontally, one can place a few drops of water on the surface and it will spread out to make a splendidly reflective layer. As I lifted this wonderful find from its dark muddy hiding place, it was quite spooky to see myself, cap-light and all, reflected in this ancient mirror that had lain below the surface for centuries. In fact, a German film crew was making a documentary for European public television down there in August 2005, and we reenacted the discovery. The cameraman was even able to capture my squiggly reflection in this mirror on film. How magical . . . like something straight out of an Indiana Jones movie!

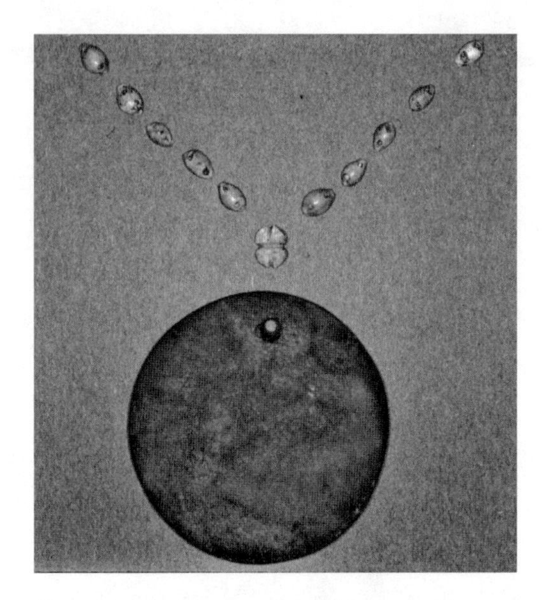

Figure 21. Ornaments from Makauwahi excavations (the arrangement shown here is purely conjectural): cowrie shells drilled lengthwise; beneath which is a carved bone bead with a single lengthwise hole and incised grooves around both axes; and, at bottom, a polished black basalt mirror drilled as a pendant.

This splendid artifact had a tiny round hole drilled near the edge, as if to wear it as a pendant (figure 21). Nearby in this layer I found a remarkable large carved-bone bead, in a style not like any that anyone in the group had seen before, and not pictured in any books that we searched. Also apparently associated were many small cowrie shells, drilled lengthwise. All of this, mounted on a finely braided cord from native olonā fiber such as the ones we found in this layer, would have made some priest one heck of a magical amulet. In this pit we also found a type of small stone rocking stool known as a *noho* in Hawaiian, and evidence for ten postholes in a pattern that could have supported a small platform. Scattered through this layer were many large chunks of charcoal, as if fires had been extinguished prematurely. There was also a complete skeleton of a pig cooked in an *imu* (earth oven) and ceremonially buried without having been eaten. What did it all mean?

LaFrance had a family story that seemed to match the evidence perfectly. She told it to us before hearing about the archaeological mysteries of the site. Her great-great-great grandfather, Keahikuni Kekauonohi, was the owner of the cave at the time of the Great Mahele (Land Court Awards) in 1848. This was when the colonial government required all native Hawaiians to register their land claims with the territorial court, or else lose their rights to it. Many, through unfamiliarity with this strange new custom or trusting the foreigners too much, found themselves divested of lands they had occupied for generations, and foreigners owning a lot of the best pieces. By this time, native Hawaiians had died in droves from European diseases, and the coastal forests had been leveled to extract sandalwood (*Santalum* spp.) for export to China.

LaFrance was the traditional Hawaiian owner of the cave, although Grove Farm Company owns it in the current legal sense. She told us a story she heard as a child from an older family member on a visit to the cave (which was still a shallow pond in those days, a half century ago).

Keahikuni was a *kahuna,* a traditional diviner, who could see the future. He sat on a wooden platform (presumably on posts set in the postholes we excavated) back in the mouth of the South Cave, where there was a dry sandbar. People would wade into or walk around the edge of the shallow pond and come to him for advice. As they asked their questions, he would kindle a small fire, then scatter the coals and look in the ashes and smoke for visions.[7]

It was gratifying to have one of those cases in which traditional lore and archaeology seem to agree, and even edify each other. But that wasn't the end of it. One of Pila's last projects was to find the cave's real name. This is something that had bothered us for more than a decade—that such a spectacular natural feature would not have its own name. It was as if the true name was some lost secret, a kind of etymological fossil waiting to be found. There were plenty of modern, prosaic names, including Limestone Quarry Cave, Grove Farm Cave, Warriors' Cave, the Sinkhole, and so forth. Pila and I had used Māhā'ulepū, for lack of a better Hawaiian name, although this term applies to the whole coastline there and the *ahupua'a,* or traditional land unit, that borders the cave on the east. The cave itself forms the boundary between this and the adjacent ahupua'a of Pa'a to the west. That is probably why the pig was buried there, as this was a ceremony practiced at the boundary (ahupua'a literally means "pig marker"). To many local people, the place was known as the Sinkhole, although I never liked that name much because it seemed unflattering to such a spectacularly beautiful feature.

Pila excavated the place's real name, apparently lost for a century or more, from one of the Lahainaluna student essays from 1885: "Kua is a fishing ground. When Maona appears above Kioea that is the lower land mark. When Makauwahi, a long cave at Paa, seems to appear at the brow of Makaihuena, that is the upper land mark."[8]

In one of his last research accomplishments, Pila had breathed life into our beloved place with a rechristening. The name Makauwahi for the caves and sinkhole at Māhā'ulepū may also have a connection to the story from LaFrance about her ancestor Keahikuni. Although spelled strangely for Hawaiian (in those days the spelling of the language had not been fully standardized), the name probably derives from *maka uahi,* which could be translated "eye smoke," "in the smoke," or "source of the smoke," perhaps referring to smoke in the eyes or seeing through smoke. Burning brush we cleared in there, we had long noticed what Keahikuni and his supplicants must have seen, too. When you build a fire in the sinkhole, especially in the long months of strong trade winds, the smoke wafts upward in slow graceful spirals, as the entire sinkhole performs like a huge chimney.

From that moment just a few years ago when all these puzzle pieces fell together, the place has had a real name. The name spread like wildfire along the "coconut wireless," as island folks call the gossip network. Within weeks of this discovery, people all over the island were calling it Makauwahi Cave.

Pila and LaFrance played big roles in solving another place-name mystery. All the tour books say that Māhā'ulepū means "falling together," or "and falling together," perhaps in reference to the second great battle fought here. In 1796, Kamehameha I, the ruler of all the other Hawaiian islands, tried for a second time to conquer Kaua'i. On the rough crossing from O'ahu, a great storm hit his fleet of 1,200 outrigger war canoes (said to have been carrying his 10,000 bravest men), sinking many and forcing others to turn back. A few made it, however, allowing the very weary occupants to sneak ashore at the mouth of Waiopili Stream by Makauwahi Cave. While the exhausted warriors were sleeping on the beach, the local militia discovered them and attacked. Those who weren't killed jumped into their canoes and started paddling back toward O'ahu, but they apparently realized in time that when they arrived, they would be executed by the king for retreating. Instead, they paddled all the way down to Kawaihae, a remote shore on the Big Island, and hid there. To this day, this story is told in two places, Kaua'i and the Kawaihae area of Hawaii, at the other end of the archipelago.

But the venerable folklorist Gabriel I says that story, though it may be true, has nothing to do with the name Māhā'ulepū. In fact, with the official macrons in place, Māhā'ulepū is not pronounced correctly for "falling together." Instead, he insisted, it actually means "lazy penis foreskin." That explains why, I realized on hearing this, Hawaiian ladies fluent in their native tongue always seemed to get embarrassed when I proudly gave the place its correct Hawaiian pronunciation.

Anyway, Mr. I wasn't sure why it was called that, but he theorized that it might have something to do with the likelihood that the men living on that coast in the old days probably spent long days out on the water fishing and came back a little shriveled. Pila and I always figured that was a poor explanation, even though we didn't have a better one. But we were intrigued by the opinion, expressed by LaFrance and other folks who knew the area's folklore, that at one time there was said to be a powerful stone, an *ule* or phallic stone, somewhere down there. Indeed, there is a stone at the front of the South Cave that some say was a "birthing stone," one of those vulva-like seats that crop up around several Hawaiian sacred sites, supposedly a place where royal births occurred. LaFrance, Pila, and others always insisted that this was not the "stone of power" at Māhā'ulepū, that there was another one, "a male one," but nobody could remember exactly where.

Well, one day we found it. LaFrance had loaned us, for a week's work, a couple of dozen young Hawaiian Americorps volunteers who worked for her in Kaua'i's Habitat for Humanity program. Normally they build low-cost housing for local families, but they were doing their "enrichment" experience, in which they participate in another community service activity with an educational component. So they were helping us clear out the exotic trees and vines that had been choking the sinkhole and wedging apart rocks of the walls with their roots. Near the end of this enjoyable and productive week, I felled a big banyan tree that was wrapped around some tall rock formations along the west wall of the sinkhole. It shook the ground with a jarring thump when it hit. After the roots were stripped off the rocks, there it was: a huge natural stone phallus, capped with a big smooth stalagmite that looked, from a certain angle, just like a flopped over, "lazy" penis foreskin. Pila, LaFrance, and I concurred: the area's long-lost namesake had returned, after hiding out in a hollow tree for decades!

It was gratifying to discover that so many old stories were in fact more or less true, and there were plenty of other examples. Many local people insisted that, although Māhā'ulepū and Pa'a are essentially uninhabited land today, hosting only "absentee" land uses such as quarrying and leaseholder agriculture, in the early nineteenth century the area was densely inhabited by Hawaiians. We suspected that early on, because we had excavated with Pat Kirch (an anthropologist at the University of California, Berkeley) a small pit on the beach near the Gillin house with nineteenth-century midden material. Both here and in the nineteenth-century layers from the adjacent cave excavations we had noted that, despite decades of contact with Europeans and Americans, the Hawaiians had continued making stone and bone tools and seemed to have few European goods. With no suitable anchorages for big ships nearby, this area had apparently remained conservatively Hawaiian long after Waimea, Wailua, Lihu'e, Hanalei, and other Kaua'i towns with relatively good harbors had become the haunts of whalers, traders, and missionaries.

Figure 22. Hiram Bingham drew this sketch of lower Māhā'ulepū Valley in his notebook in 1824. At least 34 human structures are visible, suggesting that the area was densely populated in the early nineteenth century. A single tree is visible in the lower left. Haupu Ridge is clearly delineated in the background, fixing the village location as the vicinity of Makauwahi Cave Reserve Unit 2. (Courtesy of Missionary Children's Museum, Honolulu)

Searching through the notebooks of Hiram Bingham, an early missionary, Pila found a remarkable thing that confirms these notions. Bingham had made a detailed sketch, dated 1824 and plainly labeled "Mahaulepu," of the landscape as it appeared then from the beach near the cave (figure 22). The distinctive outline of Ha'upu, the ridge of mountains framing the valley's eastern side, is shown clearly in the background, thus fixing the exact location of the landscape depicted. In this view, instead of one house as today (the Gillin house), at least 34 man-made structures, very traditional looking huts with thatched roofs, can be counted. Perhaps significantly, there is only one tree in the picture, a palm.

Pila and I also found and copied many nineteenth-century maps of the area. A comprehensive and well-preserved collection of them is housed at the Grove Farm Museum in Lihu'e. The more detailed ones, such as a map from 1886, show the sinkhole and, just outside it, Kapunakea Pond, which

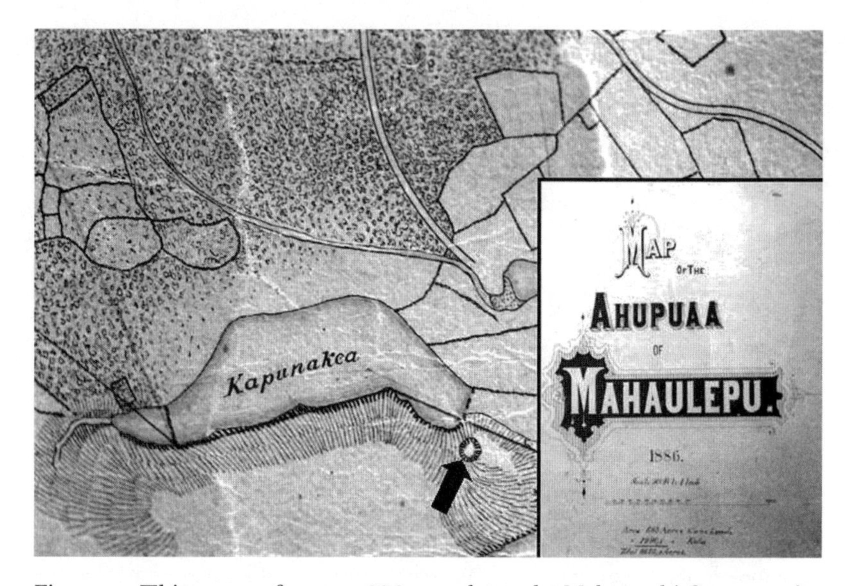

Figure 23. This excerpt from an 1886 map shows the Makauwahi Cave area, including the large brackish pond (Kapunakea) that formerly existed on the site, as well as the sinkhole itself (indicated by arrow added by the author). Inset not to scale. (Courtesy of Grove Farm Museum)

we had heard lots about from local folklorists (figure 23). The maps show that the pond, situated on the landward north side of the sinkhole, formed a large crescent adjacent to the ground-level entrance to the North Cave. This pond, as previously mentioned, was probably connected tenuously to the sea over a sand bar, and was partially drained by channelization in the 1950s by Elbert Gillin. Excavation at the North Cave entrance showed that Kapunakea was probably integral with the pond inside the sinkhole until it was drained. This would have allowed fish to swim into the brackish pond from the ocean on a very high tide, find their way through the small cave entrance to the inner, fresher pond, and most likely never find their way out. No wonder, as I dig through the sinkhole's soft sediments with my bare hands, I so often get punctured by a fish spine from centuries or even millennia ago. The cave was, among other things, a big natural fish trap.

We have an actual photograph of Kapunakea Pond (figure 24). The

Figure 24. Photograph of the area that includes the cave, looking southwest, probably taken between 1890 and 1920, showing a limestone escarpment (Kiahikea Cliff) with the entrance to the North Cave near the left end of the picture. Much of the hill on the right side of the picture has since been quarried away. Note the closely grazed appearance of the landscape, with very little grass or woody vegetation. The strip of water across the middle ground, up against the escarpment, is the former Kapunakea Pond. (Photo courtesy of the Bernice P. Bishop Museum)

great sheet of water stretches across the background, snuggled up against the limestone cliff that contains the North Cave entrance. In the foreground, there is a lone person among lava boulders and bare soil, with scarcely any living vegetation in sight. Pila thought this photographic treasure from the Bishop Museum Archives was taken between 1890 and 1920. By that time, most of the local Hawaiians had long since died of European diseases or moved away, all the remaining trees had been cleared for more than a half century in the pursuit of logs for export to China in the sandalwood trade, and feral livestock had proliferated. Goats, sheep, donkeys, horses, cows, and pigs had apparently eaten everything in sight, setting the long-stable dune fields back in motion, as soil-holding roots gave way to bleached barren sand. You may recall that this reactivated dune sand, and the feral livestock's bones, form a distinct layer in the sinkhole.

There are also old stories from later times. In an obscure memoir about life in the area during the ensuing plantation days, *Return to Mahaulepu,* Charles Katsumu Tanimoto refers to an "underground lake" (our sinkhole, undoubtedly) that served in the early twentieth century as a secret "Chinese opium den."[9] Since reading this account, we had wondered if we would someday dig up a fossil opium hookah or something like that. Digging a big hole in the surficial clay one day a couple of years ago to plant a native tree in the sinkhole, Lida and I did find a peculiar old glass bottle of a type we had seen in museums that often contained laudanum, a form of opium once easily obtained in drug stores. Other "countercultural" events in the cave's twentieth-century history included, according to local folklore, a hippie commune that used the place around the 1960s for drug-induced mind-altering experiences.

Apparently this group took the rainbow as its symbol, probably reflecting the more ancient symbolism of the area's most famous petroglyph. "Rainbow man," depicted in books on the subject, the Kaua'i Museum in Lihu'e, local restaurant walls, and T-shirts, shows a manly stick figure holding up the arc of a rainbow. He is seldom seen anymore, as the stones on the beach containing this and other wonderful petroglyphs, including a detailed picture that looks like a European sailing ship but with Polynesian-style *lauhala* (woven mat) sails, are normally buried under the sand in the ocean's swash line. Every few years, usually during stormy weather, these petroglyphs appear on the beach for a few weeks, until the sand builds over them again. Mrs. Gillin, whose front yard hosted these wonderful, mysteriously appearing images, said it was really bad luck to see them, because every time they showed themselves over the years there had been a hurricane or a tsunami.

All kinds of people, with every imaginable sort of religious belief or spiritual connection, have turned up at the cave over the years. I guess this is to be expected, as the place is so spectacularly strange looking that one would have to be made of wood not to feel a certain atmosphere there. Many visitors use the adjective "sacred" to refer to this place. One can make out on the walls of the North Cave a couple of scary faces, where someone in years past used what appears to be chalk and charcoal to enhance facelike sculptures in the rock. The more conspicuous of these, one our kids called "Mr. Funny Face," actually frightens some people with its stony countenance. At night it seems to glow a little in a flashlight beam.

One of the most controversial spiritual activities to have taken place there actually overlapped our early years in the cave. A local witch coven used to meet in there in the dark of night, drawing their pentacles and nine-foot circles in the sand, burning black candles, and the like. I didn't mind this too much, having seen and survived plenty of genuinely spooky things in my years in Africa and Madagascar. Then we were required by federal and state authorities back in 1997 to put up a strong gate and lock people out when we weren't there to watch the place, to protect the burials in the back. Somebody kept breaking into the gate, and one of my local "spies" informed me that it was probably "the witches." Don't get me wrong, now, some of my best friends are witches, as they say, so no slight intended. But I did a little sleuthing myself, and found out how to get a message to them. My message was that if they knew somebody who felt they needed to get in there badly enough to damage our gate, I would just give them a key if they asked for one.

That apparently ruined it for whoever it was breaking in, as I figured it might. In any case, vandalism of the gate subsided, and so far no one has ever asked for a key for any ceremonial purpose.

TWELVE

Irrigating the Future

IT HAD BEEN SLOWLY DAWNING on me for several years that my interest in the past was not separable from my interest in the future. Seeing so much extinction more or less firsthand can turn you into a kind of missionary for conservation, I suppose. In truth, I have been a shameless advocate for the environment at least since age fifteen, when I taught Nature and Bird Study merit badges at Boy Scout Camp Uwharrie in Guilford County, North Carolina.

It sounds a little corny, but a few years ago I got interested in the possibility that the Poor Man's Time Machine has a forward gear as well. To put it more seriously, I wanted to examine the likelihood that information from the fossil record and all our other past sources could be used fairly explicitly to design and implement ecological and cultural restoration. Situated in its dramatic coastal setting, with land still partly supporting native plants and animals, what better place than Makauwahi Cave to try out this idea? Perhaps here on Kaua'i, could we create a kind of public park dedicated to combining study and reverence for the past with creating a better future for the area's natural and cultural resources? Visitors could participate in ongoing excavations, but they could also view and assist with a wide variety of restoration projects that we might dream up for the surrounding landscapes.

We were talking about this concept as early as 1997, as we unearthed a surprising variety of fossil seeds, showing what a diverse plant community had thrived here for millennia. I found an appropriate slogan to use in my slide talks. It was borrowed from the French philosopher Henri Bergson: "The present drains the past to irrigate the future."

Several real or imagined obstacles stood in the way of just removing the nonnative vegetation and planting back the original forest. First, we didn't own the land; we just had permission from Grove Farm and LaFrance Kapaka-Arboleda to do research there. Second, we didn't really know if these plants (and potentially some animals) that once thrived here could manage in the changed modern circumstances. And third, where would we get the huge amount of labor and considerable sum of money to make all this happen?

In the last year or so of his life, Pila and I began meeting with the Grove Farm officials to discuss the possibility of obtaining the property surrounding the cave. In talks with Allan Smith, Dave Pratt, and Mark Hubbard in the boardroom of their office in Puhi, it became clear that there was no possibility to buy the property, because it was not for sale. But they would entertain giving us some kind of lease on the property at no charge if we could find an institution that would cover the cost of $1,000,000 in liability insurance coverage. We embarked on long conversations (a discussion that lasted more than two years, actually) with representatives of Kaua'i Community College and the National Tropical Botanical Garden, to see if somebody would take it.

As for whether the species we wanted to reintroduce could bear up under the changed modern circumstances, we believed most could. Our exhaustive studies showed that the biggest changes for these species were probably human overexploitation, which we felt we could manage, landscape transformations, some of which we thought were reversible, and biological invasions—that was the worrisome one. But we figured that if people worked at controlling invasive plants and animals in the vicinity (this would be pretty easy inside the sinkhole at least, we guessed), former natives might have a chance. Quite a few natives had persisted just outside the sinkhole, shedding their pollen and seeds into sediments before, during, and after human arrival, right up to the present. Perhaps with the right care and timing, it might be possible to enrich this latent community with some of the rarest species, locally disappeared elements still holding their own at least tenuously in the island's remote interior or elsewhere in the Hawaiian Islands (figure 25).

At this point, Pila did something highly symbolic that made all the

Figure 25a. The sinkhole as it appeared in 1997, after volunteers assisted the researchers to clear away most of the tangle of aggressive vines and trees. A large pit is under way on the left (southeast) side, complete with shade tent for screeners and a water pump and hose for excavation below the water table. Compare to color plates showing the sinkhole as it appeared in 1996 and 2009.

difference, in hindsight. Well before the negotiations and paperwork were finished—years before—Pila went ahead and started planting native trees down there. His first efforts were the now large and lovely grove of kou trees (*Cordia subcordata*) on the sinkhole rim near the overlook. With the help of our good friend Ed Sills, a local businessman and chef, and various KCC students, Pila got a start on planting around the turn of the millennium. To be honest, I was skeptical at first, even though we knew from the fossil seed capsules and pollen grains in the sediments that kou trees were growing there for thousands of years before the Polynesians arrived (contrary to most books, which say they were brought by the Hawaiians). I was worried that restoration wouldn't work in this loose, infertile sand amid tough, aggressive woody invaders like *haole koa* and *kiawe.*

Figure 25b. The sinkhole in 2004, two years after the first native plantings inside.

Visitation had been down in the years since we stopped digging quite so actively and were not finding so many new things. I wasn't sure that there would be enough local volunteers around to look after a bunch of native trees, watering them through that first long summer dry season, and maybe the second. But Pila was sure this was the best way to motivate the community to take an interest in this new, future-oriented phase of the cave project that was emerging.

"Just go ahead and plant 'em, Burney," I remember him saying. "Don't worry—people will feel sorry for them and do the watering." His faith in community spirit worked. Since that time, volunteers, both a small cadre of dedicated locals as well as several thousand students and visitors, have helped Lida and me to manage the spectacular 17 acres surrounding our time machine. Sadly, Pila didn't live to see it come to full fruition. It is fitting, though, that he planted the first of the thousands of native plants we would ultimately grow here, and that they have lived and done so well, eventually poking through the holes I opened in the forest canopy of gradually disappearing nonnative trees that I was cutting out, one by one.

What we really wanted to do, something we had all been talking about since 1997 at least, was to relandscape the nearly one acre of "sunken gardens" inside the sinkhole. With help from NTBG nursery manager Bob Nishek, NTBG field botanist Ken Wood, arborist Mark Query, and a host of other brave, hardworking souls, we had been removing the remaining banyans and Java plums, and wresting the open space inside the sinkhole from the annoying aggressive Guinea grass (*Panicum maximum*) and a host of persistent nonnative weeds. One native plant survived in the sinkhole from the long years of competition with the new invaders. That tough legacy plant, our only definite residual from the original flora still inside, was a single individual of that wonderful sprawling vinelike shrub *maiapilo,* the native Hawaiian capers (*Capparis sandwichiana*). Somebody noticed it when we were pulling all the nonnative vines out of the rocks on the west wall. Over the years, with the blessing of a whole community's "aloha," that vine has continued to expand over its part of the wall, putting out puffy white blossoms almost every night of the year in hopes that some pollinating moth will come by.

Getting all those weeds out of the sinkhole was not going to be easy, we knew. Everyone involved agreed that using pesticides, including herbicides like Roundup and Garlon, to take out the undesirables was a bad idea, and our state and federal advisers concurred. If we put chemicals out inside the cave, they are trapped in a small, enclosed system, a kind of giant terrarium. That terrarium becomes an aquarium just below the ground surface, because groundwater stands today at approximately the level of the pond that was in there for millennia. Any chemical added to the system, if it gets into that water, would eventually wind up, we believe, in the little pond that outcrops on the cave floor today at the very back of the South Cave. That's the area we call the Troglobite Room. Dr. Frank Howarth of the Bishop Museum discovered in the 1970s that our cave and a few lava tubes near the adjacent town of Koloa contain some remarkable living invertebrates specialized for cave darkness (figure 26). The primary consumers in this always totally dark ecosystem are a little shrimplike amphipod, *Spelaeor-*

chestia koloana, and a cave pillbug, the isopod *Hawaiioscia* cf. *rotundata.* They eat much smaller organisms that feed on detritus that washes into the cave during infrequent storms, and on the sugars and other nutrients exuded by any roots that make it into the cave passages while searching for water. These tiny creatures have no pigmentation—their exoskeletons are a clear, whitish film over the body that scarcely hides the organs below. The blind amphipod uses its antennae and front legs like a whole battery of white canes.

Sounds peaceful, right? Actually, even though these little blind arthropods have probably been in this and nearby caves for millennia—maybe even tens or hundreds of millennia—there is a predator to beware. The Kaua'i blind cave wolf spider (*Adelocosa anops*) also lives in the vicinity of the Troglobite Room, and in a subset of the nearby lava tubes with the blind amphipod. This big eyeless spider, generally pale beige to almost pure white in color, is apparently descended from the large, fast, hairy "big-eyed spiders" found on the surface throughout the islands. So our top carnivore, a spider too big to hide under a quarter, feels its way around with long delicate legs, and, perhaps aided by excellent hearing, catches and eats the amphipods. That's right, one federally listed endangered species makes its living eating another endangered species. Nature can be really shameless that way sometimes.

This bizarre lightless ecosystem was not something we wanted to mess around with. These little creatures are bound to be rare, because they are island-bound at more than one level. They live on a single small volcanic island in the most isolated archipelago in the world. They live only in one small part of that island, the few square kilometers from the town of Koloa down to the coast at Māhā'ulepū. In that small part, they occupy only a few rare geological features (lava tubes and a limestone cave), and only a small portion of the caves (areas in total darkness with minimal airflow and near 100 percent humidity). Their numbers are estimated in the dozens to single digits at each site, and they are thought to be, like troglobites (cave-restricted organisms) studied in places as far-flung as Alabama and Rumania, highly sensitive to any chemical changes in their environment.

Figure 26a. The Kaua'i cave amphipod (*Spelaeorchestia koloana*), one of the eyeless cave creatures found in the darkness of Makauwahi Cave. (Photo by Wendy McDowell)

Figure 26b. The Kaua'i blind cave wolf spider (*Adelocosa anops*), the top carnivore in the Makauwahi Cave troglobitic ecosystem. It is completely eyeless, unlike its daylight cousin, the big-eyed spider. As a result our spider is often referred to oxymoronically as the No-eyed Big-eyed Spider. (Photo by Gordon Smith)

Figure 26c. A unique photo of an immature Kaua'i blind cave wolf spider eating its fellow endangered species, the Kaua'i cave amphipod. (USFWS Photo by Michelle Clark)

So, thanks to our peculiar little neighbors, the troglobites, we were determined to convert the sinkhole and its rim from invasive plants to native plants by the "softest" methods possible. We wanted to avoid not only chemicals but any sudden drastic changes in the soil, water, nutrient, and light regimes. The basic plan was to turn the tables species-wise, but do it as gently as possible, little by little.

Dr. Adam Asquith, an entomologist who has worked with the U.S. Fish and Wildlife Service, the Waipa Farmers' Cooperative, Sea Grant, and Kaua'i Community College, brought a Youth Conservation Corps group down to the cave for a tour and work contribution to the cave project in early July 2002. I had hoped to be there, but our schedules eventually failed to mesh; I had to leave the island the day before. I had arranged, through the National Tropical Botanical Garden, for some native trees, shrubs, and groundcovers for them to plant in the sinkhole, recently bereft of many of the more aggressive nonnatives and all ready to plant. These first few dozen plants were of species that we believed from the fossil record were predominant on the landscape surrounding the cave before humans. These included native loulu palms, rare trees and shrubs, and some more common species, including ones that survive today in the dunes and rocks near the sinkhole.

Some of our dedicated volunteers, especially members of the Malama Maha'ulepu group that has advocated conservation in the valley for years, watered these new plantings, which included some on the outside of the sinkhole near the overlook as well. I would not see the results until I returned many months later. For years I had been coming back to a depressingly overgrown sinkhole after being away for some months, then spending several days to a week of precious research time hacking it all down so that we could work. This time, instead of just the usual head-high Guinea grass and castor bean, I expected to see small struggling native plants, overtopped by head-high Guinea grass and castor bean.

That is not what I saw. To this day, I can remember vividly the scene that greeted me as I stepped up to the rim of the sinkhole at the overlook that day. The new plants were thriving and holding their own on the site, thanks to the efforts of Jeri DiPietro, Rob Culbertson, Dave Crawshaw, Ed Sills, and others. Pila was right—if we plant them, somebody will take care of them.

Standing on that rock ledge looking down into the sinkhole, I was seeing more than endangered native greenery. At that moment I felt, more strongly than ever before, that I was looking at the past and the future simultaneously, tenuously joined by good will in the present.

From that day on, my time down at the cave would always be partly committed to taking care of this place we all loved so much. Plants needed to be watered, weeds had to be whacked or (better still) pulled up, and decisions would need to be made about what to plant where. I really wanted to get a lease on the property nailed down and in general spend more time working out there on both research and restoration. Meanwhile, I had very successful field seasons in Madagascar during the summers of 2003 and 2004, was promoted to full professor at Fordham, was granted a Fordham faculty fellowship for another sabbatical, and juggled these things with many trips to Kaua'i to continue the research, plant and care for native plants, promote the idea of a Makauwahi Cave Reserve, and try to broker that lease.

It emerged over the course of the meetings with representatives of Grove Farm, the landowners, that what they might be able to offer was technically called a "license" for the property around the cave. This is similar to the arrangements they make with local farmers, tenant farmers in a sense, who cultivate portions of Grove Farm land but do not live on it. An actual lease on these properties, according to local and state law, would require subdivision approval, which would mean lots and lots of paperwork and delay.

What I worked out with Grove Farm was that I would demarcate on a very accurate current aerial photo the outlines of the parcels we would like to take under license. They wanted me to break it into two parcels—actually, a core area that we felt we needed in order to minimally protect the cave environment, which encompassed the footprint of the cave, including land on top of known subterranean passages, and a small buffer around the cave land. In addition, I had requested (and this request was the slightly trickier part) that the reserve we wanted to set up would also need some acreage across the stream, including small wetlands, a large abandoned cane field, and a road right-of-way up to the present Grove Farm gate on the main road to Māhā'ulepū Beach, the Gillin house, and Kawailoa Bay. The draft license we came up with included both parcels, and the time they would give us on the first license was five years. There were, of course, the usual several pages of legal language, such as bailout clauses, but I did manage to get them to include a provision that the license would be offered to us again, upon expiration, as a ten-year license. Lida and I figured if we were going to stick our necks out on this crazy project, we would need five years to see if it was going to work, and a guarantee that the incipient park would have another ten years at least to become the permanent institution we would like to see there.

The liability insurance, a million dollars' worth of coverage with Grove Farm as "other insured," was a nagging detail. That much coverage was going to cost around $3,000 per year. Alas, Lida and I are not rich, and we knew that it would be awkward or even impossible to write a grant proposal to fund an insurance bill. The obvious solution was for a full-fledged local institution to adopt the project and extend their insurance coverage to it. Pila was sure we could eventually work it out with his beloved Kauaʻi Community College, and he did his best to make it official. After he died, I tried to fill his shoes in those negotiations, but it was always apparent that the University of Hawaii system might have trouble justifying it, as it is self-insured and underfunded. As time dragged on and this insurance seemed the only obstacle to getting our arrangements for the property completed with Grove Farm, I began working on another strategy.

Even before KCC began to give serious consideration to "adopting" the cave project, the National Tropical Botanical Garden, located nearby in Kalaheo, had seemed like a logical partner in this effort. That was in the late 1990s, when Bill Klein was director. That hope had waned after he passed away and the emphasis shifted at NTBG under Dr. Paul Alan Cox. Now, Cox was leaving, and the new acting director was one of Kauaʻi's most respected local conservationists, our friend Charles "Chipper" Wichman, Jr. While I was on sabbatical, one of my projects on Kauaʻi was to develop a restoration plan, with U.S. Fish and Wildlife support, for the Lāwaʻi-kai section of the lower part of Allerton Garden, managed by NTBG. Members of the NTBG staff and I developed a restoration plan for three lovely acres (1.2 ha) in front of and alongside the historic Allerton house, using our fossil results from the Makauwahi Cave, Lāwaʻi-kai's own estuary, and other sites along the south shore of Kauaʻi.

It made sense at this time to rebroach the idea of NTBG adopting the cave property. In the spring of 2004, back at Fordham teaching after my sabbatical, I made many short trips to Kauaʻi (long weekends between my university lectures) to participate in various collaborations. I used this precious time also to press hard to nail down the cave license and the issue of institutional affiliation, even showing the place to a group of NTBG trustees and fellows. I promoted the idea at their spring meeting at the Lawai Gardens, and gained the recommendation of a blue-ribbon panel of conservation experts. This group met at the NTBG headquarters (and toured the cave, of course) to chart the future course of NTBG's Conservation Department. In their report on the conference, titled "Strategic

Directions for the Twenty-first Century: Envisioning the Conservation Potential of the National Tropical Botanical Garden," one of their official recommendations was as follows: "We enthusiastically endorse NTBG's assuming the lease on the Makauwahi site, and recognize the direct relevance of findings from Quaternary paleobotany and paleontology to the definitions of the restoration efforts that are taking place throughout the State, an increasing number of them sponsored by NTBG. Makauwahi is perhaps the most extraordinary site in the Hawaiian Archipelago for such findings, and furthermore is a vibrant center of community interest, volunteerism, and restoration activity. It will be a great asset for many aspects of the Garden's programs."[1]

Nice. Although admittedly, as the compiler and editor of this report, I wrote parts of it, I believe Dr. Peter Raven, director of the Missouri Botanical Gardens, crafted this particularly nice bit of advertising for the cave. As one of the world's best-known botanists and one of the smartest and funniest people I ever met, it was great to have his participation in the NTBG Conservation Summit, and especially gratifying that he liked our cave project so much.

But when I signed the formal agreement for the cave property on June 15, 2004, giving us five years and the option of another ten, NTBG was not the new owner. Neither was KCC. NTBG decided they couldn't risk taking on any more property right now without an endowment (which we didn't have) to support its upkeep, and KCC concluded that it wanted to remain involved in the education programs but couldn't pay the insurance.

We had to do something or risk losing this opportunity forever. There was no guarantee that, if we didn't do this in 2004, the offer would still be there in future years. Grove Farm has been working on its own strategic plan for its nearly 50,000 acres (20,234 ha), owned by Steve Case, formerly of AOL-Time Warner, and there was a lot of talk in the community about what would ultimately happen to the property. The Malama Maha'ulepu organization had been actively campaigning on the political front for years to get the valley and coastline under some form of public protection, undoubtedly state or federal. On the other hand, the success of the Hyatt Regency Hotel, on the same end of Poipu Beach as Māhā'ulepū, naturally has kept the idea alive with Grove Farm's managers that someday they might build a hotel, or an upscale housing development, somewhere in the valley.

After a lot of agonizing, Lida and I decided to just take the plunge ourselves. For the time being, Makauwahi Cave Reserve would just have to be, on the books at least, a mom-and-pop operation. Grove Farm was willing to take this route. So we bought some insurance, presented the certificate, and signed. From that day, we like to tell people, "We don't really own the cave, but the cave surely owns us."

In the summer of 2004, things had to change rapidly for us. For months, even years, we had been moving toward the idea of living on Kaua'i. After fifteen years at Fordham, I was ready for a change. I wanted to stay affiliated with Fordham if possible, to continue working with graduate students and keep our state-of-the-art paleoecology and conservation biology laboratory there in operation to serve our own research program and that of our collaborators. I also hoped to continue teaching my summer courses at Fordham, perhaps offering a field-school component on Kaua'i. But I wanted in general to teach less, and "do" more. I really enjoyed teaching, but I wanted to see if it was possible to interrupt an extinction event, if only on a small scale. I figured that, in my mid-fifties, perhaps with luck there was still time to do one more crazy new thing, professionally speaking.

Specifically, run the Conservation Department at the National Tropical Botanical Garden on Kaua'i, just a few miles to the west of our beloved cave. Chipper Wichman had offered me the position of director of conservation. I agreed to take it on a one-year trial basis, meaning that either party could bail out after one year, commencing from September 2004. That way, I could return to Fordham if things didn't work out, or even if things there somehow looked better from that perspective after the passage of a year (not likely).

So we sold our house in Croton Falls, New York, moved most of our stuff to Lida's family house in Carteret County, North Carolina, and took the rest to Kaua'i. The garden had, in negotiating a salary arrangement with me, thrown in the use of a small cottage on a delightful ridgetop overlooking NTBG's portion of Lawai Valley. It would be tight, but since Mara was now a grownup staying in New York for her job and graduate education, our number would shrink to three. Alec enrolled in Island School as a junior. I worked part of the summer as a volunteer for NTBG, to get the lay of the land, went off to Madagascar for five weeks and did some exciting new excavations and cave explorations, and started to work at NTBG in September. By that time, some grant proposals I had written, aimed at jump-starting new conservation initiatives at NTBG, had been funded. Lida and I had also obtained support for planting native species in the Makauwahi Cave Reserve, through the Wildlife Habitat Incentives Program of the Natural Resource Conservation Service, U.S. Department of Agriculture. I became a conservation administrator who put in some weekend and after-work time as a volunteer at the Makauwahi Cave Reserve, and Lida became a full-time native-plant tenant-farmer.

THIRTEEN

The Tour

OKAY, HERE YOU ARE AT MAKAUWAHI Cave Reserve (figure 27). You probably emailed or called Lida and made an appointment, or you came for the regular Sunday morning "Open Cave" tours. Anyway, here we are in our little parking lot behind CJM stables, on the western edge of our lease property. This is the worst-looking, most disturbed part of the place, Management Unit 6. When the quarry was opened in the 1950s, the surface spoil on this side was shoved over onto this area, creating little man-made dunes. People come here to camp, fish, and watch the offshore excitement, because this area looks off the sea cliff onto the rocky shoals and steep underwater drop-off of the Gillin's Beach area of Māhā'ulepū. Whales, especially mothers with calves, are often surprisingly close here in late winter, and it's also a good place for seeing spinner dolphins, green sea turtles, and monk seals any time of year; the seals bask on the beach and tend their pups there. Just about any day with decent wind and surf (or even without) one can see surfing, wind surfing, and especially kite surfing folks taking full advantage of this windy southernmost bit of Kaua'i.

On the property's highest spot, not far from our one maintained campsite, you can see all of the *ahupua'a* of Māhā'ulepū, and Pa'a behind you (figure 28). It is a great sweep of ocean, shoal, shore, beach vegetation, lowland irrigated agriculture, limestone quarry, and pasture. Pila loved this spot so much that we have come to refer to it as Pila's Point. He always said to classes and other visitors, with a grand sweep of the arm, that we should be growing native and Polynesian plants all over this valley, but particularly there ought to be some native loulu palms (*Pritchardia*) on this spot, instead of just ironwood and other nonnative scrub. Now there is a

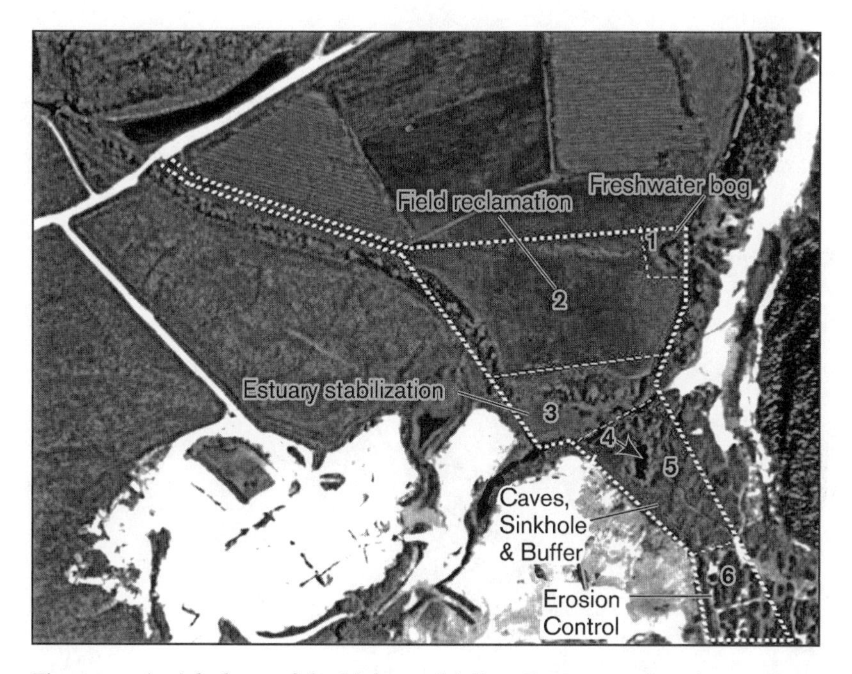

Figure 27. Aerial photo of the Makauwahi Cave Reserve and environs. Outer dashed line indicates the boundaries of the reserve and access right-of-way. Inner dashed lines delineate management units, each with different methods, goals, and challenges for ecological restoration. Unit 1: A small freshwater bog originally containing two native sedges, now enriched with additional species of native wetland plants. Unit 2: An abandoned agricultural field originally containing only one native plant species, now enriched with nearly 100 native and Polynesian-introduced plant species. Unit 3: Estuary stabilization project, including erosion control measures and establishment of native riparian species. Unit 4: Demonstration garden inside sinkhole, initially containing one native plant species, now featuring 23 species of native and Polynesian plants, most of them particularly well represented as fossils in the sediments. Unit 5: Sinkhole rim and headprint of the cave passages, initially containing six native plant species, enriched with about 20 more. Unit 6: Erosion control area, where mitigation of vehicle damage to the sand dunes and invasive plants led to the recovery of five native dune species, now enriched by additional ones. Large white area at left is the limestone quarry that operates adjacent to the cave property.

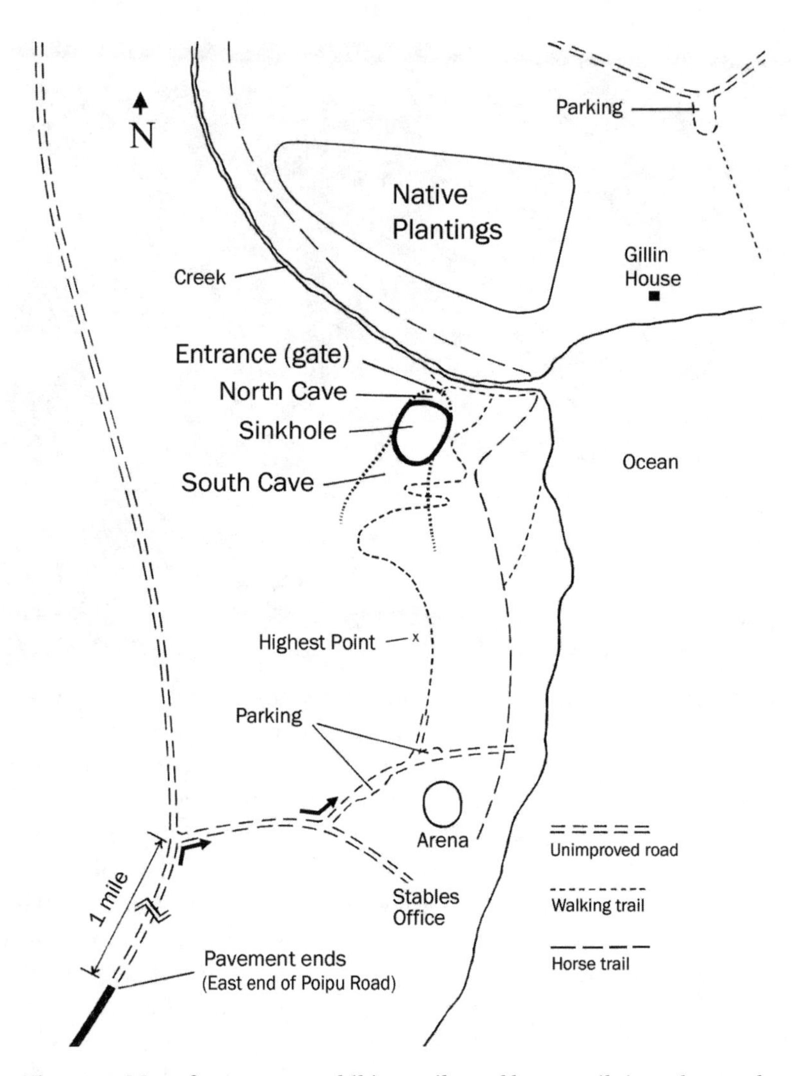

Figure 28. Map of access routes, hiking trails, and horse trails in and around Makauwahi Cave Reserve. Arrows indicate car access off the paved east end of Poipu Beach Road. (Drawing courtesy of Mel Gabel)

whole cluster of *Pritchardia aylmer-robinsonii* on the hillside, watered by the 1,000-gallon (3,785 liter) irrigation tank we installed, until somebody stole it recently. Now it's connected to our automated irrigation system from across the stream. We have a gravity-fed supply for automated watering all over the property on both sides of the stream. Off in the distance, looking more or less eastward, you see the rim of the sinkhole, our field of native trees beyond, the Gillin house perched on the beach—the only house in the scene—and Adam Killermann's leased corn fields beyond.

Make your way down this bit of ridgetop trail, taking care not to fall into the abandoned quarry section on the left, and you wind up in Area 5C and D, the "headprint" of the South Cave. This area was an absolute thicket of thorny kiawe (*Prosopis pallida*) and haole koa (*Leucaena leucocephala*) before Lida and groups of interns and volunteers cleared this vegetation, killing the stumps the hard but clean way, either pulling them out or smothering the sprouters with several thicknesses of weed mat covered with rocks. The herbicide Garlon-4 will kill the stumps a lot faster, of course, but any chemicals applied here will almost certainly find their way into the cave passages and troglobitic organisms below. Remember, that food chain starts with exudates on plant roots and organic debris that washes in, so let's keep it "organic."

The most spectacular of the native plant species growing here, in my opinion, is not the sweet-smelling naio (*Myoporum sandwicense*) or the kou, but that scrambling shrub-vine, maiapilo, or Hawaiian capers (*Capparis sandwichiana*). Its filamentous white blooms open nearly every night, fading in the heat of the day. The fleshy orange fruits smell like rotten mushrooms, attracting ants and other insects in droves. The flat dark seeds, about the size and shape of a small tick, are produced in copious quantity on the many maiapilo we have planted here. We collected the original seeds from a small population of these rare plants nearby, and one now huge specimen that was growing in the wall of the sinkhole when we originally cleared the exotic vegetation in 1997.

On the edge of the sinkhole just ahead, a metal creature is perched, looking like a Martian probe. That's the weather station and brain center of our "cyberinfrastructure" program, an elaborate high-tech network of devices we are deploying over the property (mostly tiny sensors that go unnoticed by visitors) to measure wind speed and direction, precipitation, solar intensity, temperature, and other useful environmental information. The data are beamed once per hour to the Intelesense Web site (www.intelesense.com), where anyone with the right software can check conditions here, either in real time or synoptic archives on various time scales from hours up to whole years. We have additional Web-accessible sensors around the property that measure the physics and chemistry of groundwater in the East Pit excavation, as well as monitoring soil moisture in the restoration sites, deep cave conditions, repeat photography cameras, and so forth. Someday we would like to have remotely pannable video cameras, so that Web visitors (and caretakers) can have a look around—it would be cool to have a seismometer that, in the event of an earthquake (like the big one we had in October 2006), would tell the cameras to pan toward the ocean and watch for an incoming tsunami.

Anyway, let's make our way down the slope to the overlook, where you have a great overview of the sinkhole itself, a natural amphitheater carved by groundwater and eroded by the elements, from eolianite, the sandy limestone here that is really a great petrified dune field from the late Pleistocene. Today it is the loulu palms that dominate the landscape inside, but that has changed dramatically over the years and will undoubtedly continue to do so (see color photographs). On your right is the great archway leading into the little North Cave, which contains the walk-in entrance. On the left is the even greater arch of the much bigger South Cave, the area we walked over when visiting with the maiapilo plants. Across the sinkhole is the huge high arch that overhangs the west wall and leads into the upper passages of the cave that we seldom visit but have thoroughly explored over the years. Now we leave it mostly to the white-tailed tropicbirds that nest up there. To its right and down a little is the great stone phallus described in legend, presumably the "lazy penis foreskin" that gave Māhā'ulepū its name, according to one version. It is remarkable that this giant cone-shaped mound, crowned with a suggestive stalagmite, had been lost for decades.

If we head down toward the stream, through a woodland that is turning slowly from nonnative to native, beginning with the thick groundcover of leadwort (*Plumbago zeylanica*) and trunks of native kou trees poking through the aging canopy of gnarled kiawe trees, we can pick up the trail that follows the stream up to a sharp, shady bend. By a big notch in the limestone *cuesta* (an erosion scarp, with no such nice word for it in English as the Spanish), you confront the first, and possibly only, moment of panic about going into this otherwise spacious cave. The entrance is a small dark triangle only about three feet high at the base of the cliff. Everyone—even national politicians and the wealthy and famous—must pass into our nether world of the cave in a submissive crouch, either crawling on the knees (we have special mats we put out for royal visits and paying customers) or, for children and small, cave-savvy adults, in a kind of low duck walk. Just inside, there is an old water-worn stalactite with "hundreds of human DNA samples already," as I always say when I lead visitors in, warning them not to unwittingly make a similar donation here with their head. I've probably made a dozen donations myself.

A few people over the years have simply backed out at this point—roughly one in 500—because the entrance looks too scary. But anyone who takes the trouble to peek inside from the vicinity of our big iron gate knows that, instead of the dank crawl heralded by such an unassuming entrance, the inner sanctum here is a glorious vista of high cave ceiling with helictites (small twisted stalactites that mantle the ceiling with their tortuous texture), framed by a huge stone archway leading to the brilliantly green interior of the sinkhole, a place of splendid light in the midst of a cave that started out promising a claustrophobic experience. The sight of what I just described is often punctuated with an "oh wow," "awesome," or "neat-o" from first-time visitors. It is even more surprising if you don't visit the overlook first. When people come from the cruise boats docked in Nawiliwili Harbor, bused there and guided by private concessions who work with us, we don't show them the sinkhole from above until *after,* so the sight has its full impact.

Inside the North Cave, I often experience time vertigo. The visitor from this point on is inside the Poor Man's Time Machine, and every bit of the surface—walls, ceiling, and especially the floor—is mantled with the artifacts of time travel. The structure itself is a great hollowed-out fossil, a cave formed in the very heart of ancient sand dunes turned to stone. The rock is perhaps 400,000 years old, cut by groundwater sometime in the past few hundred thousand, mantled over tens of millennia by flowstone that has formed not just stalactites and stalagmites but columns, draperies, false floors, and other bizarre speleothems from the recrystallization of calcium carbonate and other minerals. Above the sediment floor, one sees that water has been back in here since the speleothems formed, eating away at their bottom sides and along the walls.

For me, the best part is buried in the floor of soft sediment beneath our feet. Directly under us, and throughout much of the sinkhole ahead, there is an absolute layer-cake pile of many kinds of interesting soft sediments—together over 33 feet (10 m) thick in many places. This material has been laid down, our radiocarbon dating shows, over the past 10,000 years. That's right, the cave rock is older than the cave passages, which are older than the speleothems, which are older than the soft sediments, which are older than us.

But that last part is not quite right. Those of us standing here, not just the oldest, either, are also represented in the sediments underfoot. The top layer, this several feet of reddish-brown silty clay, includes the twentieth and what there is so far of the twenty-first century. A thin new layer was deposited in some parts of the cave by a flood in 2006, after I started writing this book. The uppermost layer contains the artifacts of the present, plastic bags and such, with preceding cultural layers signaled as one digs down through one's own time, by Styrofoam, Bakelite, and cellophane artifacts. Preservation is remarkable in this layer as in those deeper, with well-preserved Polaroid Land camera disposable film backs, with a vague trace of an image on the tear-off, in the 1960s and 1970s layer (That's right, folks, this place is pretty so let's take a picture and throw down the litter it produces . . .). I get a strange sense of time warp sometimes when I bring up an artifact from our field seasons of a decade ago, or something I remember from my grandmother's kitchen as a child.

Well, that's not what has made this place rather famous as a site for artifacts and fossils, but my point is that the kind of fidelity this unusual type of site has extends from the remote past right up to the present. Fossil localities that are highly remarkable for either their diversity or quality of preservation, or both, are called *lagerstätten* in German, but again English fails us. This time-transgressive lagerstätten seems to extend from the immediate present back with few interruptions for thousands of years, all through colonial and Polynesian times and back to that remote landscape that evolved before any human ever saw it.

Seeing this sinkhole, or looking at a diagram of it—neither experience quite prepares you for the idea that this was all a big cave passage that got so undermined that it collapsed into itself, but that is almost certainly what happened. The dry cave floor that existed here 10,000 years ago was rudely interrupted one day when the sea came rolling in, after being over 400 feet (120 m) lower for tens of millennia during the last Ice Age, when so much seawater was tied up in glacial ice. As this ice melted and sea level correspondingly rose around the world, a point was finally reached about 7,000 years ago when sea level was getting close to the present height. At least in this local case, close enough to flood the contemporary cave floor, roughly 25 feet (7.6 m) lower than present sea level at the site. The cave floor became a giant, mostly dark tide pool, with a layer of sand and marine shells laid down for us to find. If parallels to other, earlier-stage sites around the world can be believed (like Pancake Rocks in New Zealand), weak places in the ceiling became giant blowholes such as ones found even today nearby at Spouting Horn, then gradually eroded to form great churning holes in the collapsing ceiling.

Once all this dramatic collapsing settled down and the cave became choked with marine and terrestrial debris, a fresh-to-brackish pond then laid down nearly 7,000 years of soft, highly fossil-rich sediments. This process continued, with occasional interruption from extreme events like tsunamis and hurricanes, right up until the times LaFrance Kapaka and her relatives came here when she was a child. For thousands of years, this lake gradually filled up with fine sediments. Everything from freshwater diatoms and pollen grains to whole skeletons of giant flightless waterfowl and bird-catching owls and hawks accumulated here, like frames in a very long, slow movie. As many creatures disappeared from the record, rats, pigs, dogs, human artifacts, and food items took their place. Only a couple of centuries after a tsunami about 400 years ago rained huge stones onto this floor (exposed here by excavation) do the first signs of Europeans turn up.

You already know all that, but this is your chance to *feel* time. Stick your hands in the bucket of mud . . . let's see . . . this one says "BAC-EP, FF40, 3.4 m." So this lovely chocolate-gray goop is about 3,000 years old. Here, drop your treasure on this quarter-inch mesh screen box, with window screen in the box below, and give it a good wash with this hose . . . that's right, stir it gently with your hands, to help the fine grains wash through to the second screen . . . okay! Just look at all these extinct snails, a half dozen or more species and hundreds of individuals in two handfuls. Oh, and there's part of a bird beak, the mandible or lower jaw. Hmm . . . Storrs or Helen could tell you for sure if they were here, but let's label the box "cf. *Chloridops*" as it is so heavily built I think it is one of the heavy-duty seed-cracking finches.

"What the heck is that?" someone asks over your shoulder, pointing to a big seed, nearly an inch across, lumpy and flattened. That's from the lonomea tree (*Sapindus oahuense*). There's one that we planted growing right in front of the North Cave archway. And here's a fragment of a *Pritchardia* fruit, a perfect match for the ones on the palms we planted here back in 2002, now huge.

And on it goes, as we pick out and sort the fish bones, wood, and so on. Some people are soon satisfied; many, especially kids and elderly people, are difficult to tear away from this basic time-travel exercise. The South Cave is a good lure, as its magnificent arch, with a great multi-ton natural keystone, promises further adventure. We have passed by various pits, one still getting larger on recent weekends, and wind up by the big squarish pit inside the South Cave entrance. This is where dear old Keahikuni, in the nineteenth century, sat on his platform and saw the future in "the eye of the smoke." (Did he see us?)

On into the back, with the flashlights on, we can see very little at first. Pause here, in this great echoing room, and get your eyes adjusted to cave darkness. It's hard not to look back at the great archway from the inside, as the dark walls frame a startling landscape vignette, the intense green of the sinkhole. If you look up at the high walls, you can see one of the few places in the entire system where a large segment of the original fabric of the rock itself is visible. Look, here is fossil weather . . . these diagonal lines are the rhythmites from past wind storms: a thin band of coarse sand, then a fine, then another coarse . . . each pair probably represents the accumulation from a day's wind cycle nearly 400,000 years ago.

Most of the rest of this great room, the largest uncollapsed room in the cave, is heavily draped with speleothems of many types. On the high ceiling are white encrustations, places where salts from above are probably still accumulating. Down the sides are elaborate draperies, including fin-like and ribbonlike ornamentation. We see here the condensation-driven speleothems, perhaps some of the youngest in the cave (thousands rather than tens of thousands of years old), that we call "cave popcorn," although really big ones can look like inverted coral or clusters of grapes.

This back part of the great room is the jumping off, or rather crawling off, area for the mazelike passages ahead. We don't take the tour into these areas, out of respect for traditional burials back there, as well as to avoid stepping on the blind cave organisms, who don't generally move all that fast and certainly don't get any warning from the flashlights. Those parts of the cave, including the "Troglobite Room" with its stream resurgence (a place where the water table outcrops on the cave floor and water flows toward the ocean), are very special places that are protected both by our gates and by the arduous muddy routes to get there.

Our exhibit panels on the way out will give you an idea of just how eyeless and white a creature can be. These little blind Gollums groping around in the dark must have a really different sense of time, if they have one at all. I guess life in cave darkness would put the usual priorities on finding food, staying out of trouble with larger creatures, and mating, just like any other life, but with a definite requirement for extreme patience and efficient energy conservation. It's not that appealing—I would rather have been the bird-catching owl, or a turtle-jawed moa-nalo. Heck, even a big burgundy-shelled *Carelia* snail had it better, I reckon. But what do I know? I'm just a flimsy human trapped in my own time . . . mostly.

Okay, let's get the questions answered and get on out of here. So far we have only toured the past. The rest of the reserve is mostly a tour of the future, if species survival, ecological restoration, and cultural rebirth are what the future is about, and we certainly hope so.

Follow the trail down to the mouth of the stream, where we will either cross on the portable bridge or wade, depending on how high the water is. On the way we pass through a really beautiful little woodland, one of the few green places on this property we don't want to change at all, the Milo Patch. Here along the mouth of the stream, this tree thought to have been brought by the Polynesians really thrives (we've never found it in prehuman sediments like kou and hala, two others said to have come with the Polynesians but now known from our fossils to have gotten here well ahead of them). Called milo in Hawaiian, *Thespesia populnea* is a valued timber tree, completely adapted to life near salt water and a kind of dryland equivalent to mangroves, which have been introduced to the Hawaiian Islands in the twentieth century. The rich dark wood of milo is featured in many beautiful carvings and bowls in the art galleries and tourist shops.

Crossing the stream, watch out for horses and horse plops. Perhaps Henry, C.J., or one of the other cowboys will entertain our visitors with an exchange as he passes with a line of tourists "seeing Kaua'i on horseback," as the entrance sign used to say (or is it *from horseback*?). Coming up the other bank, we get a great view of Ha'upu Ridge, essentially the same one that Hiram Bingham sketched in his notebook along with all the houses that used to be here.

Just beyond, however, the landscape beckons us to other times, both before and after humans cleared the site. A new forest, our Management Unit 2 (everybody calls it "Lida's Field of Dreams") is flourishing there, thousands of native Hawaiian and Polynesian plants, about 100 species of them, some quite rare. This is the largest single restoration project on the property, and one of the largest examples you will see anywhere in the Hawaiian Islands of a diverse native forest that has been grown completely from scratch in an old crop field.

It is also where most of our tours end up. We can rest here in the shade of trees we have known since they were seeds. For many of these species, this is the first time they have grown here in perhaps a thousand years, as if returning from a millennial journey.

So, I hope everybody remembered to wear work clothes. If you need gloves, we can lend you some from that white bucket by the corner of the shed. We hope you brought your hat, but there are a few spares for those who didn't. Grab one of those little trees in the gallon-sized pots, please, and follow me down Row 21 to the bottom of the hill. Lida has some tools set up down there already, I think. She will show you our preferred method for planting the trees. Thanks for coming to Makauwahi Cave Reserve.

FOURTEEN

Right Here, Right Now

"He ali'i ka 'Aina, he kauā ke kanaka"
[The land is a chief, the people its servants]
—Hawaiian proverb

AFTER THREE DECADES OF RESEARCH concerning how species become endangered and eventually go extinct, we had finally come to the point of daring to try something new. If we can understand how species go extinct from human carelessness, can we also develop programs to stop this senseless waste of biodiversity? All conservation is based on the premise that the answer to this question is a general "yes." In this specific case, can we find clues in the past to make a better future for some Hawaiian plant species, and can we do it fast enough and on a sufficiently large scale?

This is pretty incredible hubris, I guess, to think that the kind of human-caused extinction catastrophe that has happened everywhere over the past 50,000 years could be stopped in its tracks out here on Kaua'i at this time. We're a long way from that, for sure. If anything, the conservation news from Kaua'i and the entire state of Hawaii is pretty dismal, has been for some time, and shows few signs of getting better.

The Hawaiian Islands are ripe for new conservation ideas for an obvious reason: the situation really is dire. The lowlands of Kaua'i, for instance, are almost entirely lacking in native vegetation and suitable habitat for endangered animals. Exotic vegetation blankets the landscape up to roughly 3,300 feet (1,000 m) elevation or more in many areas, with some notable but highly threatened exceptions. Thousands of acres are covered with invasive alien plants; some, such as rat-berry (*Rhodomyrtus tomentosa*), Guinea grass, and haole koa, form essentially single-species stands that successfully exclude most natives and even many other alien invasive species. No native passerine birds are regularly seen below about 3,300 feet, as a result of the presence of the introduced mosquito *Culex quinquefasciatus*. Most endemic land snails, including the large, colorful *Carelia* species, are believed to be extinct. Giant flightless waterfowl, the original meso-herbivore community of the islands, have been extinct for centuries, and endemic flying waterfowl, such as the nene and the Koloa duck (*Anas wyvilliana*), are on the federal endangered species list.[1] And don't forget—Hawaii has 0.2 percent of the land area of the United States, and 43 percent of its officially listed endangered plant species. About half of the native Hawaiian plants are on this list, or should be. Depressed yet?

Formal status reviews of each endangered species are just beginning for the long list of Hawaiian plants, but results from 21 species evaluated by the Conservation Department of the National Tropical Botanical Garden in 2006 under contract from the U.S. Fish and Wildlife Service show the magnitude of the challenge in microcosm: all these species, save perhaps one or two, have shown declines since their official listing under the Endangered Species Act. Most have declined not only in total number but in number of populations. These species were selected on the basis of administrative considerations (their turn on a 5-year update cycle), not because they were perceived to be at greater risk than others on the list. In nearly all cases, the cause of decline is not known with certainty, but the observed or inferred negative influences are the usual litany—the same factors indicated in the fossil record of island extinctions in the human period, including overexploitation, habitat modification, and invasive species.[2]

When things look bad, define your terms, I always say. Anyway, before I say that conservation is very nearly a failed enterprise out here in the islands, let me clarify what forms conservation can take and the challenges of each strategy. From its earliest days, the emerging multidisciplinary enterprise called conservation biology has recognized two ways of doing conservation: *in situ* and *ex situ*.

In situ in this context refers to conservation efforts applied to species in a preexisting wild condition in their current range; *ex situ* is conservation efforts based on intensively human-controlled environments, such as botanical gardens and zoos, genetic banks, and propagation facilities.

The community of conservation professionals and dedicated volunteers in Hawaii is sizeable, and many energetic projects, both *in situ* and *ex situ,* are under way. *In situ* conservation will always be the front line of defense for rare species. But most rare species in Hawaii are in very remote places, where they must fend for themselves most of the time against invasive ungulates, competing weeds, diseases, insects, rare-plant poachers. When they are visited by botanists and restorationists, there is always the risk that well-meaning folks will inadvertently bring along new invaders. Finally, many rare species are holding out where they are in the wild today, paleoecology suggests, not so much because that is their favorite patch of habitat but more likely because the spot is too steep or otherwise inaccessible to pigs, deer, maybe even goats.

Ex situ offers the advantage of the protection afforded by a botanical garden or other rescue facility, but the plants are on full life-support in most cases and no longer part of an evolving natural environment. They are, like that tiger in a cage, alive but not really part of the living world. Besides these philosophical drawbacks to *ex situ* conservation, there is also the hard economic reality. Space on these arks is costly, and therefore limited. At NTBG we are currently trying to grow and plant out about 100 at-risk Hawaiian species, but what about the other 400?

Leading conservationists in Hawaii have called not only for a greater effort but for a more focused, creative approach, based on the best science available and promoting research. Two ideas in particular have attracted attention at regional conservation meetings since about 2004 as possible breakthroughs in addressing the overwhelming conservation challenge. It could be argued that these are simply two sides of the same coin. On one side is the emerging practice, now familiar, of using local paleoecological, archaeological, and ethnohistorical sources to develop restoration plans and propose reintroductions for managed areas; on the other is a new strategy, *inter situ* conservation, that in effect bridges the gaps between *in situ* and *ex situ* approaches.[3] *Inter situ* conservation seeks to reintroduce species to locations outside their current range but within their recent past range. In some cases, closest living relatives or ecological surrogates may be substituted for globally extinct species that are considered essential to the function of the target ecosystem.[4] These two ideas have proved to be highly synergistic in that, for instance, paleoecological findings are increasingly used to support proposals to create new populations in the late prehistoric and early historical range of declining species.

The term *inter situ* has been used in conservation for more than a decade, and the notion got a good airing in the Hawaiian conservation community in July 2007, at the annual Hawaii Conservation Conference in Honolulu. I organized a symposium for the meeting titled "Rewilding, Island Style: New Ideas in Inter Situ Conservation." This session featured six speakers, all NTBG affiliated and including Lida and me, talking about all aspects of this idea as applied locally, including not just the cave project but a host of similar *inter situ* projects on NTBG properties, as well as those of government agencies and private landowners collaborating with us. Michael Soulé, one of the fathers of conservation biology in the United States and one of my favorite people in all of science, led the discussion afterward. He was there to give the keynote address for the conference, in which he asked "Is Island Conservation Fundamentally Different from Continental Conservation?" and concluded that, primarily because the invasion challenge is more acute on remote islands, it probably is different in most respects. He led our session's discussion, a very positive interaction with the large audience of conservationists from throughout Hawaii, with his usual skill, insight, and humor. It also served to officially inaugurate the word *rewilding* to the local lexicon. As you will discover in Chapter 15, this word has taken on a kind of potency around the globe in recent years, getting people excited.

In practice, most of these projects involve a variable mix of horticultural and agricultural techniques, in which reintroduced species are subsidized for a time, but husbandry is eventually, often gradually, withdrawn. Familiar examples would be "soft release" techniques for reintroduced animals, and temporary irrigation systems, periodic soil amendments, and weeding for reintroduced plants. Although the site may resemble a barnyard or cropfield initially, the ultimate goal is usually a phased withdrawal of most direct care from the recolonizers (ungulate exclusion fences being a major exception) and a hope for reproduction and recruitment success. A key advantage with most *inter situ* projects over more remote *in situ* locations is that greater accessibility and a lack of jurisdictional complications make it generally more feasible to correct and continue addressing the challenges that resulted in a species' decline in the first place.

Paleoecology has played a variety of supporting roles in this effort on Kaua'i. First, studies of past ecosystems have shown scientists and the public the full magnitude of extinction losses and ecological transformations that have come in the wake of the human disturbance, reinforcing the sense of urgency. Second, information of this type has provided direct scientific justification for efforts to implement corrective measures, such as feral ungulate management and exclusion, increased agricultural inspection controls over incoming materials that might introduce new invasions, and protection of archaeological and historical sites. Third, and perhaps most important from the standpoint of ecological theory, paleoecological findings have revealed some surprising details about the formerly much wider ranges of now rare plants and animals in prehuman Kaua'i and subsequent changes in their environments. Finally, paleoecology, environmental history, and ethnographic information about landscapes and species give interpretive and educational programs a better sense of place by providing, in addition, a sense of time in a place.

The main reason this *inter situ* idea and its big globetrotting cousin, rewilding, are grabbing so much attention, aside from the radical idea of reconstructing ecosystems almost from scratch as a way to save rare species, is the potential bigness of it. If we can grow rare plants anyplace we choose within their late prehistoric range, then we can often grow them where it is convenient to take care of them. Many of the last *in situ* locations for our rarest plants on Kaua'i are reachable only by helicopter or arduous climbing. Often they are visited only once per year or less, and "visiting" may consist of dangling off a sheer cliff on a rope. If you can grow these plants efficiently in places like Lida's "Field of Dreams" in Management Unit 2 of Makauwahi Cave Reserve, doing a lot of the work with a tractor, you can consider implementing conservation on a sufficiently large and affordable scale to add a huge number of individuals to the species pool.

Standard procedure in all these projects is to conduct baseline studies, document carefully all restoration treatments and the genetic pedigree of the plant stock used, and monitor the results (figure 29). Sites vary in the intensity of monitoring, but most projects include repeat photography stations, digital mapping of plant locations, periodic vegetation sampling, and establishment of manual or automated devices for measuring local weather conditions, water quality, and soil conditions. NTBG, the Waipa Farmers' Cooperative, and Makauwahi Cave Reserve have established, in collaboration with University of Hawaii researchers and Intelesense Technologies at Stanford University, wireless data loggers that report weather and water parameters to a Web-accessible database. As we mention in the tour, implementation of a much-expanded network is under way at Makauwahi Cave, with grant support through the University of Hawaii, to implement other technologies such as automated time-lapse photography of the plantings, soil loggers to record ground conditions, and sensors to monitor the delicate deep-cave environment. We hope someday to also have video cameras that can be panned remotely. A seismograph will measure the shake from earthquakes like the big one in 2006 and conceivably even direct the cameras to turn toward the ocean to record any possible (heaven forbid!) approaching tsunamis. Tauber traps (pollen collectors) at Makauwahi Cave Reserve are used to continuously monitor the site's airborne particulates in a range of environments. It will be interesting to see if one day, after more extensive restoration, the pollen spectra collected from the air are similar to the ones in the prehuman sediment record that inspired the restorations in the first place. Now that would be cool.

One of the most interesting future challenges for *inter situ* restoration on Kaua'i, and in other places, centers on the role of genetics in decision making. The pioneering work of Michael Soulé and his colleagues set the standard early on for including genetic considerations in conservation planning.[5] Clearly, genetic variation is generally a good thing in conservation, as natural selection needs something to select from to find solutions to environmental challenges. Variations among plants in disease resistance,

Figure 29. Repeat photography of a panorama of Makauwahi Cave Restoration Unit 2 over a span of less than three years shows the remarkable growth of native vegetation on the site. (Photos by Alec Burney)

resilience to climatic challenges, and other key traits help guarantee that some members of a population can survive in a world of change.

The possibilities introduced by *inter situ* restoration, however, pose important questions and promise a rich harvest of results from large-scale genetic experimentation with reintroduced populations. Whereas this type of work has progressed with animals to the point of genetically enriching small relict populations as the key to survival, as with the well-studied examples from panthers and wolves in North America, this idea is still in its infancy among plant conservationists.[6]

For instance, conventional thinking has supported the idea that new plant populations, and tiny relict populations, should be kept genetically "pure" by using only stock from the nearest *in situ* population or populations. But in the kinds of situations typical for many of the rarest Hawaiian plants, the nearest "population" may consist of only one or a very few individuals. What to do? Try to create new populations using only a tiny fraction of the potential genetic variation available, by sticking to this logic, or create new populations infused with as many genomes as possible assigned to this species? For example, if a new *inter situ* population is to be created for a species represented by a nearby *in situ* population of two individuals, but three more individuals exist elsewhere on the island, should the new population be started from two, or five, founders? Even the extremely conservative restorationist would probably opt for capturing the additional variation represented by the bigger set. But what if the other three are on another island? Or, what if one population consists of twenty individuals, and the other, thirty? Clearly, these are tough questions for which we have only limited data.

Paleoecology, however, can provide a surprising amount of insight. For instance, several species used in restorations at Makauwahi Cave and NTBG have widely disjunct populations today, none near the site. But throughout the Holocene, until the advent of humans (as recently as a thousand years ago), these species probably extended across the area between these disjunct populations, based on their occurrence in various fossil sites, including the cave vicinity.[7] In other words, present distributions are merely a recent human artifact, so rejoining the human-separated populations would be the most prudent solution. Similarly, since many *inter situ* populations are in rather different habitats from any of the modern disjuncts (which are usually on steep cliffs or other rough terrain in the interior, not necessarily because they prefer this habitat but because this is the only safe haven from goats and pigs), arguments in favor of preserving narrow genotypes in new populations give way to the need to provide maximum variation so that appropriate phenotypes for the "new" habitat can emerge.

Overall, the *inter situ* strategy will in reality be one of adaptive management, as scientists, horticulturists, and volunteer groups collaborate to learn their way through this local and highly challenging version of the global biodiversity crisis. Well before the quantitative results of most formal experiments are harvested, native-plant growers will have found ecological combinations that work for a given site through systematic trial and error, common sense, and occasional hints from the records of the past. At Makauwahi, for instance, the first appearance in the record of Polynesian-introduced Pacific rats coincides with the sudden disappearance of nutritious, thin-shelled seeds of the native *Pritchardia* palms, but thick-shelled seeds of species of *Zanthoxylum* persist well into Polynesian times. This hinted strongly to us that native palms could be reestablished only with effective rat control, and the project members have followed this "advice" from the fossil ancestors of the plants being reintroduced through the establishment of rat-free zones and, where this is not feasible, placing wire cages over the *Pritchardia* inflorescences to protect the developing seeds.

So far, it is safe to say only that each site poses unique challenges, such that horticultural generalizations are less useful than site-specific knowledge. For instance, it is becoming clear that on drier sites—with less than perhaps 60 inches (1,500 mm) of precipitation per year—certain native groundcover species can effectively compete with many invasive weeds if given a head start under the right circumstances. Wetter sites, on the other hand, have so far been manageable only with labor-intensive hand removal or mowing—or a larger scale of long-term herbicide use than many professionals would prefer and a large segment of the local public will condone for a "natural" area. Hand weeding, for instance, is the most environmentally benign and reliably effective weed control in these wetter areas, but only small areas can be maintained this way with currently available labor, as demonstrated repeatedly in the Limahuli Lower Valley Preserve restoration projects. On drier sites such as the Makauwahi Cave Unit 2 field reclamation, where annual rainfall is only about 30 inches (762 mm) and with high year-to-year variability, repeated cultivation by tractor, followed by several thorough hand-weeding sessions and establishment of dry-adapted native groundcover plants, has so far satisfactorily controlled invasive plants without any herbicide use. Experiments slated for the near future will test systematically the use of geese and perhaps other domestic animals for weed control, as has been practiced so effectively with tortoises on the Mascarene Islands of the Indian Ocean. Perhaps tortoises could help us control weeds someday (several people in the islands have them already), serving as an easily fenced, controllable ecological surrogate for our extinct turtle-jawed moa-nalo.

One interesting question for ecologists, conservationists, and resource economists has often been posed by visitors. What are the ultimate fates and purposes of these *inter situ* sites on Kaua'i? An obvious motivation is to provide additional habitat to newly created or enhanced populations of native plants, and the native animals that use them. This is a given, and the initial results, even for some of the rarest single-island endemics, are exciting and gratifying, especially in the face of the continued grim crisis of biodiversity loss in Hawaii. In both theory and practice, however, other uses are emerging, some quite obviously utilitarian.

On sites reclaimed from fallow agricultural fields initially lacking natives, such as Makauwahi Unit 2, the planting design and the management approach have been geared from the inception to creating a place that will be managed as an extractive reserve. Plants established here, and their offspring, are providing nursery stock for other restorations and gardens. Since 2007, NTBG's intern groups have assisted field botanist Natalia Tangalin of NTBG to collect thousands of native plant seeds from Lida's "Field of Dreams." An assessment of the value of seeds collected there in 2008 for other restoration projects, based on Internet sources and local experts, put these materials at $7,000. Other materials that are being extracted include plant products useful to cultural practitioners, woodcrafters, and herbalists; taxonomic and genetic study material; and other biological resources to be managed sustainably. One project getting under way at several NTBG sites and Makauwahi, headed by Tamara Wong, a University of Hawaii grad student, involves production of a native plant used in lei making, the coveted maile vine (*Alyxia stellata*). It is currently at risk of overharvesting in the island's native forests for the production of fragrant leis especially popular for high school graduations (the state imports literally tons of maile from the Cook Islands each year for this purpose).

In another innovative application, native grasses, shrubs, and other groundcover plants are grown in patches at Makauwahi that serve as miniature "hayfields." When the natives set seed, they are scythe-mown, and the straw is gathered and spread around native trees and shrubs as a mulch and seed source in hopes of establishing native groundcovers with minimal labor input.

In addition to native plants, some restoration sites contain areas that target not prehuman Kaua'i, but early Polynesian times, as a reference system for restoration. At Lāwa'i-kai in Allerton Garden, for instance, the stated goal of the restoration project was to reconstruct the plant community, inferred from fossil pollen and seed studies conducted nearby, as it might have appeared shortly after Polynesian arrival. Thus, some restorations feature plants believed from the fossil and archaeological record, and generally confirmed by traditional lore, to have come in the double-hulled canoes of the founders of the Hawaiian people. Of special interest in this regard are projects that focus on *ex situ* or *inter situ* preservation of old Hawaiian cultivars of breadfruit, banana, coconut, sweet potatoes, and that key Polynesian staple crop, taro or kalo (*Colocasia esculenta*).

At the Limahuli and Kahanu Gardens of NTBG, and Makauwahi Unit 2, an explicit goal is to reestablish not only native and Polynesian plants, but also old agricultural varieties from the Polynesian period. In the midst of restored native forest at Makauwahi, for instance, one comes upon a patch containing ancient Hawaiian varieties of sweet potato, which Lida has found make an excellent groundcover that can compete with noxious nonnative weeds beneath native trees. Historical documents, ethnographic accounts, and even the rare find of a centuries-old yam tuber in the sediments of Makauwahi Cave support the idea that such root crops were grown in these dry areas in precontact Hawaii, probably in this very field.[8]

The southern section of Makauwahi Unit 2 demonstrates another unusual technique: an incipient overstory of native trees and shrubs has an understory of vegetable garden. To demonstrate the possibility of growing native plants compatibly with commercial or subsistence farming, a variety of edible noninvasive crops are produced beneath and between native woody species. We like to joke that this may be the only place in the world where you might see an overstory of extremely rare *Munroidendron racemosum* trees, and an understory of . . . watermelons!

Of course no discussion of any grand ecological scheme these days is complete without more than just a grateful nod to ecotourism. National Tropical Botanical Garden restoration sites, such as Lāwaʻi-kai and the Lāwaʻi Stream restorations, are part of regular guided tours conducted by organization staffers and trained volunteer docents for thousands of visitors each year. Makauwahi Cave and NTBG each offer special tours for the passengers on the numerous cruise vessels that dock almost daily at Nawiliwili Harbor in Lihue. Waipa Farmers' Cooperative features, in addition to a weekly farmers' market, special events like the annual Taro Festival that highlight Polynesian and native plants being grown and used in restoration projects. At Makauwahi Cave, tours are available any day by appointment, or simply by showing up any Sunday morning for "Open Cave." Independent operators offer tours that include a view of portions of the Cave Reserve on foot, by boat, or on horseback. Iliahi House, Grove Farm's remote, upscale conference center perched on the slope of the dramatic Kilohana Crater, features, in its front yard, nearly 5 acres (2 ha) of native plant restorations on the Makauwahi Unit 2 model, with one of the most spectacular panoramas in all of eastern Kauaʻi spread out for backdrop. The restoration plan developed by NTBG staff for Grove Farm Company, drawing its plant list from fossil pollen and seed studies from southern and eastern Kauaʻi, calls for up to 30 acres (12 ha) of native plant restoration here over the next five years.[9]

These projects have the right mix of aesthetic appeal and informal edification to attract a portion of Hawaii's massive tourist trade, but they also stand up well as high-quality formal outdoor education for everything from kindergarten to postdoctoral research and adult education programs. The NTBG sponsors clubs in all of the island's high schools—called Junior

Restoration Teams—that channel interested students into a multisession training program up to eight days per year, in which these young applied ecologists do all facets of the work associated with the restoration projects described here. More than 650 local students participated in this program in 2006, with more in subsequent years. By the end of the school year, they have worked through a complete syllabus of field sessions providing on-the-job training in invasive species control, native plant propagation, cultural site restoration, stream management, Geographic Information Systems—even paleoecology. JRT members, like other school groups, come to understand fully the meaning of biodiversity loss when they excavate and catalogue subfossil plants and animals in Makauwahi Cave, then go out and help reintroduce many of the same species to their former haunts.

Someday, but probably not soon, it really may be possible to have an *inter situ* restoration site that will generate pollen spectra similar to those in the late Holocene before humans, or in the early days of Hawaiian settlement. At this stage we are far from that. Despite the efforts of the many devoted Kauaians working to bring native species back from the brink of extinction, most *inter situ* restorations will in reality be small islands where past meets future in a presently vast sea of species that have, in many cases, been on the island for only a matter of decades. Nevertheless, the history of conservation here may show that conservationists bold enough to surf the current extinction wave, despite their possibly naive idealism, managed to bring some species back from the brink of extinction by buying enough time for these species and communities to benefit from more elegant solutions that scientists may yet discover. Until then, restoration paleoecology may be one of the best ways to help island evolution stay on the track to a better future, by encouraging conservationists to cast the widest safety net conceivable for species that are otherwise slipping away.

Not everybody agrees with these assertions. Over the past few years, as Lida and I have promoted the idea of using the past to guide restoration activity, we have occasionally been reminded of the German philosopher Arthur Schopenhauer's three stages in the acceptance of the truth.

"All truth passes through three stages," he wrote. "First, it is ridiculed. Second, it is violently opposed. Third, it is accepted as being self-evident."

Comments more on the negative side we have heard regarding our *inter situ* ideas include:

> Stage 1: ". . . it's crazy . . . it won't work . . . it's not that simple . . . more or less pie in the sky."
> Stage 2: ". . . it's unnatural . . . it's probably illegal . . . it distracts people from serious conservation . . . it doesn't look good."
> Stage 3: ". . . what's the big deal . . . everybody knew this all along . . . this is what we've all been doing for years . . . of course it will work, so what?"

All we can hope for is to get everybody to Stage 3. Actually, I think we're almost there. The key part to work out, as always, is how to get enough people and enough resources together to do something.

In the same sense that all politics is local, all conservation is, too. Whether any conservation scheme is going to work in the long run primarily depends not on how much funding or high-level political support it has initially but on the extent to which it is supported by the local community where it is embedded. Lida and I have been saying that for 30 years or more, but the Makauwahi Cave experience has solidified the notion for us.

To some extent, this is just the result of the way things evolved down at the cave, starting with some scientific research that grew and grew. Meanwhile everyone who passed by, from local fishermen, surfers, and cowboys to visitors off the cruise boats whose total lifetime experience of Kaua'i will be measured in hours, became a part of the cave experience. Their stories, advice, and even sweat became a part of the place, as they pitched in with screening of the hundreds of cubic meters of sediment I've lifted out of those pits in labeled buckets.

As I've said, Pila Kikuchi really started the next stage about a decade ago, by planting those first few kou trees up near the sinkhole overlook and challenging the community to keep them watered. Ever since, whoever has showed up there has been part of both the historical and the futuristic mission of the place. Although Lida deserves most of the credit for keeping things running smoothly day to day and staying ahead of the weeds—and both of us have planted our share of trees down there—it is truly staggering to look around that landscape and try to imagine which of the many hundreds of pairs of hands that have been lent to the project actually planted each of the thousands of natives now growing here.

Of course the National Tropical Botanical Garden has been a pioneer in the practice of doing restoration with large volunteer groups. At Limahuli, work on the native and Polynesian plant gardens featured there, as well as in the nearly 600 acres of nature preserve in the lower valley, has been done in large measure by community volunteers, especially visiting youth organizations.

The thing we would all like to see more of in particular, though, is large-scale participation by people who live on Kaua‘i. NTBG's Na Lima Kokua (the Helping Hands) organization is made up mostly of dozens of devoted retirees who have moved to Kaua‘i in recent years or come here seasonally ("snow birds") to escape the North American winters. Only a few persons who have spent most or all their lives on Kaua‘i are active volunteers at NTBG, at least until recently, even though locals make up the bulk of the organization's employees. The trick is to generate community outreach to do ecological restoration, and vice versa. Watch out—this is a kind of big thinking that calls for big action.

To stem the tide of extinction and environmental degradation, ways must be found to fully engage public support for the needed large-scale measures. The needs of native plant conservation have been much discussed in professional and public venues, and nearly every source agrees that all conservation efforts must operate at large scales to be biologically meaningful, and that no program on behalf of a rare species has much chance of success in the long run without broad-scale public support. Our efforts to prevent the extinction of some of Hawaii's rarest plant species must move forward on several fronts, and at scales large enough to make a difference for the species at risk.

If this *inter situ* approach can be used successfully to grow rare native plants in an abandoned agricultural field that was choked with half a decade of noxious weed growth (Lida's Field of Dreams), then there is a germ of a larger idea here. Literally thousands of acres of abandoned farmland are sitting idle on Kaua'i and throughout the Hawaiian Islands, growing the same dozen or so species of noxious invasive weeds. Could the local public team up with the big companies and wealthy individuals who own these lands to "rewild" some bits of it with native plants? Can they most efficiently be grown on the *inter situ* Makauwahi Unit 2 model? And just what is this model?

In Management Unit 2 at Makauwahi Cave, we have taken an abandoned field used for decades for growing sugar cane and corn, and converted it into a reforestation project (figure 30). The place was more than head-high with grass, weeds, and gnarly shrubs when we used a big tractor mower to shred the field in 2005, then disc it several times. Timing this with the onset of dry weather killed most of the weedy cover, although the seed bank of invasive weeds, and one common native, 'uhaloa (*Waltheria indica*), came on with a vengeance afterward. This onslaught was kept away from the natives without resort to any herbicide by planting the introduced natives in rows, like a "crop" composed of a mixture of dozens of native trees, shrubs, palms, grasses, sedges, and herbs, and using a giant rotary tiller mounted behind a two-wheel modular tractor (my favorite big toy) to agitate the intervening spaces and dry out the resprouting weeds. Adjacent to the plants, hoeing and hand weeding during the establishment phase gave the newcomers a chance to outstrip the competing weeds.

After establishment of this grid of plants and successful weed control, we begin filling in the paths between the rows with additional species of partial-sun and shade-loving plants that are compatible with the emerging canopy and understory. The result can be some surprisingly natural looking plant groupings and vegetative layers—an ad hoc system for establishing a lot of plants quickly and tending them efficiently that we have come to refer to as "digital landscaping." Like pixels in a computer file or digital photograph, each plant has a pair of coordinates that identify its location

Figure 30a. Kite-based aerial photomosaic of Makauwahi Cave Restoration Unit 2, taken in July 2008. Note how the diverse crowns of the many native species gradually interlock in such a way that rows are obscured.

Figure 30b. Kite-based aerial photograph of a portion of the Grove Farm Ecological Restoration at Iliahi House, on the flank of Kilohana Crater in eastern Kaua'i, taken in July 2008. These plants were all set out by the Scouts of Kaua'i on the same day, but native tree species grow at very different initial rates, we have found. (Photos by Matthew J. Bell)

in the grid. This greatly facilitates relocating any particular plant from our database, as no GPS locations are required—each row has a permanent tag number, and each plant a unique numerical position in the row. With such a simple, robust system, volunteers and school groups can assist in maintaining the inventory and plant health records with only a minimal introduction.

To our surprise, the local Farm Bureau of western Kaua'i awarded us its Water Conservationist of the Year plaque in 2007 for the cheap, simple, temporary irrigation system we set up on this unit. This involves pumping water from a nearby irrigation ditch, filtering and tanking it, then delivering it to each native plant with a setup modeled after the low-pressure drip irrigation systems developed on Israeli kibbutz farms many years ago. With a homespun but effective system of line taps and manifolds we can water our more than 2,000 plants in this field with a flick of the wrist. Also we can tailor separate areas to differing water regimes. After getting through two dry seasons, the plants are gradually weaned off the supplemental water. Every few months we "fertigate" by dripping dilute fertilizer and, at other times, mycorrhizal cocktails (beneficial root fungi) on the plants through the automated irrigation system.

So the prototype Poor Man's Time Machine works, but can it be mass produced? Can we put together a sort of kit, starting with some indirect knowledge of plant distributions in the past, but also including a prefabricated garden shed with all necessary gardening tools, a few acres of farmland, plenty of irrigation water, and abundant human energy? That last one could be the trickiest, we knew, as it is not that hard to turn out a big group to plant native trees, but the before and after—clearing land to plant, and maintaining the weed situation after planting—are a harder sell. Volunteers will show up for these activities, often motivated by curiosity. These are dreadfully hard kinds of field labor, though, the kind of work done largely by migrant workers in North American agricultural systems these days. Most people get their curiosity satisfied well before the work is done.

For the first challenge—clearing land and preparing the soil—we were satisfied that a lot of the work could be done hundreds of times faster with farm equipment, if your restoration site is a plowable field. So manual labor can be cut at that stage, if you pick a place flat and open enough.

It's easy to recruit folks to plant. Even politicians are sometimes available for this kind of work. It has a kind of reassuring aesthetic, planting trees, that falls right in there with vegetable gardening, caring for pets and livestock, and raising children. I guess it's something to do with love, actually, a kind of caring about nurturing other lives that reciprocates in subtle and profound ways. The remaining great challenge, how to care for plant lives already created, with routine weeding maintenance, pest surveillance and control, water and soil management, will always be a challenge and can scarcely be turned over entirely to machines at our present state of technol-

Figure 31. Scouts from throughout Kaua'i converged on the flank of Kilohana Crater, November 18, 2006, to plant the first two acres of the Grove Farm Ecological Restoration at Iliahi House.

ogy. Somebody has, in terms of responsibility, to "own" these plants, at least until they can go it on their own a few years hence.

Okay, so we can do a lot more on behalf of native plants by helping others throughout the community to grow native plants, and somehow instilling a sense of "ownership" in those who planted them—or somebody. It was time to test this notion.

November 18, 2006, was a day in point: almost 200 Boy Scouts and Girl Scouts from throughout the island converged on a spectacular site on the flank of Kilohana Crater in eastern Kaua'i (figure 31). State and county leaders officiated at an opening ceremony, and then the Scouts and their leaders, with supervision from NTBG employees, planted nearly 800 native trees and shrubs on two acres (0.8 ha) of the nearly 50,000 belonging to Grove Farm Company. After their weekend Camporee on the site, the

Scouts have returned on a rotating schedule in which each group tends the restoration site one Saturday every two months.

This site, next to the plantation-era Caleb Burns house at Iliahi, has since hosted another event utilizing 300 seventh graders from Chiefess Kamakahelei Middle School, who each planted at least one native tree and, like the Scouts, participated in a hike up the rim of the crater, as well as cultural activities and educational exercises related to plants. This has now become a regular community activity, with plant-out days for school groups as young as fourth grade, and daylong events for entire schools.

This all came about because the officials of Grove Farm Company decided, after touring Makauwahi Cave Reserve Management Unit 2, that they would like to work with us to do another restoration project somewhere on their vast land "like that field." Several months later, an ad hoc advisory committee (Brian Yamamoto of Kaua'i Community College, Lida, and me plus several leaders in Kaua'i's Scouting movement) assisted Grove Farm to envision the idea. We looked at several possible locations before settling on one of Kaua'i's most beautiful spots, on the eastern flank of the island's largest crater.

From Iliahi House on a clear day, which is most days, you can see half of Kaua'i, at least. Off to the southwest is Ha'upu Ridge, lording it over Māhā'ulepū, Makauwahi Cave, Poipu, and NTBG's Lawai Valley. Sweeping counterclockwise, there's Nawiliwili Harbor, Lihue, the county seat, and the airport. On around, the inland back side of Sleeping Giant (Nounou Mountain), Kapa'a and the power plant, then Bette Midler's property at Kaiwaihau, another of our coring sites. Beyond there, heading up to the north, the great jagged peaks at Anahola, and a sweep around to the broad level back of Makaleha Mountain, and over our left shoulder the vast high interior of Wai'ale'ale and Kawaikini, with Kahili Peak's long ridge leading back down to the lowlands fronting on Māhā'ulepū. There, I just turned you all the way around the island. Feel dizzy?

Lida, Brian, and I worked with Kathy Muneno, a television announcer from Honolulu, and one of Grove Farm's new vice presidents, Marissa Sandblom, from Kaua'i's west side, to put this project together with the island's Scout leaders. I completed a quick study of soils, topography, irrigation water, and existing vegetation, and wrote up a five-year plan to convert much of Iliahi House's big front and side yards into 30 acres (12 ha) of native vegetation. Thinking big, or at least *bigger.* Lori Terry-Bender, NTBG's GIS coordinator, made some nice maps, Lida and Brian helped us formulate the goals and lay out the strategy, Lida did most of the shopping, and Bob Nishek, NTBG's nursery manager, helped me move 1,000 fairly large plants to the site in a rented box truck. Lida and I erected a prefab garden shed and stocked it with tools, and then came the day, November 18, 2006. Scout Camporee. *This is going to be either really great, or really awful,* I was thinking the whole time.

Well, it was really great. Every politician based on the island, from the mayor and the former mayor to state and national legislators, helped plant the first tree, after a blessing by a respected Hawaiian elder, Aletha Kaohi. Scouts were divided into three still huge groups, roughly 50–75 in each, and each had two hours rotating on three activities. Dr. Lee Niedengarde, a local leader of the Order of the Arrow service fraternity of the Boy Scouts of America, conducted guided hikes up the crater, following a trail guide I wrote and coached him on. He taught the kids examples of three categories of plants on Kaua'i: native, Polynesian, and invasive. For the latter, there are hundreds of acres here of almost exclusively rat-berry, or downy rose-myrtle, known locally as Isenberg bush (*Rhodomyrtus tomentosa*). This is 99 percent of the plants we cleared to make our "Field of Dreams II," the first two acres of *inter situ* restoration at Iliahi. At the top, they peered over the rim into the vast, Ngorongoro-like oval crater, and heard from Lee about the Burneys' coring adventure down there back in 1992, when they drilled into the sediments at the bottom of the crater and found fossils of plants from before, during, and after the last Ice Age, including plants they would be planting today.

The second group rotation was local artisans and cultural practitioners demonstrating plant-based traditional skills, including woodcarving, basketry, and lei making. The third group was taking its two-hour turn at rewilding the hillside. Lida and seven team captains, each with their own row, supervised a half dozen or more kids in each planting crew.

Since that day, we have done quite a few "really great or really awful" events like this, and even more are planned for the next few months. Most projects involve large local groups that are making, whether they fully realize it or not at the time, some sort of long-term commitment to the idea of interrupting an extinction catastrophe, if possible.

FIFTEEN

Finding a Future in the Past

JUST HOW FAR CAN WE GO WITH this idea of using the past to guide the future? I would maintain that we already do so a lot more than is generally acknowledged, and more of this type of reasoning in landscape management is likely to be on the way. It's just good, plain old common sense. This is what humans do all the time in real life, isn't it? Other species have this capacity to various degrees as well. The basic process we use every moment in time, to shape or just ride out the present, is to refer to the past and try to visualize future alternative outcomes. This is self evident—one of Schopenhauer's Stage 3 truths.

But the theme here has of course been more specific and seemingly less philosophical than that—some of my collaborators and I have put a lot of research time in the past decade or more into finding out specific information about a particular landscape from fossils, artifacts, history, and oral traditions. We are now applying this information as best we can to designing ecological and cultural restoration projects using these perspectives from the past more explicitly than has generally been the case, although all good restoration efforts must look to the past to some degree.

All this is a harmless academic backwater, you are thinking. A halfway useful way this guy has found to weave together a bunch of adventurous world travel experiences and speculation about the distant past. Up to this point that is to some degree the case, I admit. But this idea of referencing the past to restore the future, extended to North America and the rest of the world, is a powerful, transformative—some even say ridiculous or insidious—concept.

It all started innocuously enough. Among the more progressive island nations of the world—New Zealand and Mauritius being two cases in point—work of this type has gone on for at least a couple of decades (figure 32). Very successful projects have been implemented to recreate, on small isolated offshore islets, prehistoric landscapes containing many rare native plants and animals, kept safely isolated from as many of the biological invaders on the adjacent mainland as possible. New Zealand has gotten so good at this practice that restorationists are now applying the techniques on the main islands, by creating fenced "islands" in the national parks.[1] Many of the species used in restoration, generally with pretty good success, were known from paleontological and paleoecological discoveries to have been significant components of the flora and fauna before humans arrived. The Kiwis are thus trying to turn back the clock about 800 years, to the time the Maori are thought to have arrived, with as many of the species that survived as possible. They set aside defensible areas with some rare species surviving in them, then remove the nonnative species, and replace them not just with the local survivors there but many others that have survived elsewhere in New Zealand and believed to have been previously more widespread. Sounds really familiar, doesn't it?

Making these kinds of major hedging efforts against extinction seems very expensive, a practice that only affluent countries such as New Zealand and the United States could afford. But less-developed island nations have been great pioneers in this work, from the Galapagos Islands to Mauritius. I had the opportunity to see some of this work firsthand on Mauritius, one of the Mascarene islands in the Indian Ocean, on a stopoff I made between Madagascar and Australia in 2004, and I was back again in 2007 on the way between Hawaii and Puerto Rico (other long stories).

Just across a gleaming blue harbor from the town of Mahebourg, Mauritius, stands an oval islet called Ile aux Aigrettes, "the Island of the Egrets." This is a nature preserve managed by the Mauritian Wildlife Foundation (MWF), which is dedicated to saving rare plants and animals of Mauritius and Rodrigues. The Mascarenes experienced major extinctions after the arrival of the first people (who happened to be Europeans) less than five centuries ago. Mauritius, well known in the rest of the world for its most

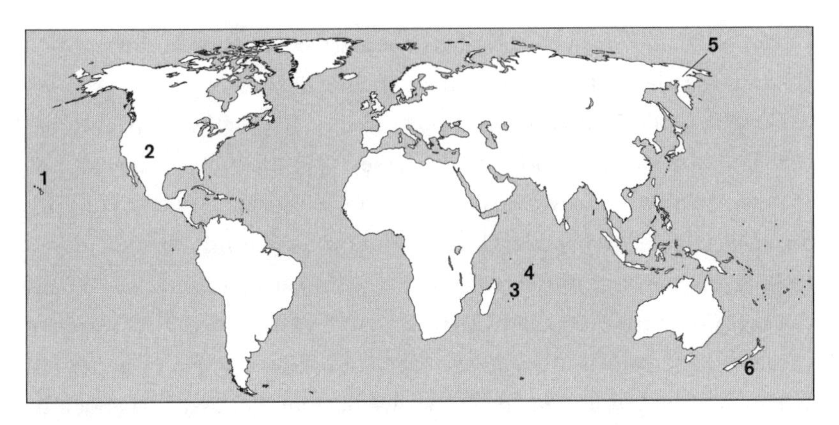

Figure 32. Locations of rewilding sites around the world: (1) Makauwahi Cave Reserve; (2) Turner Ladder Ranch, near Truth or Consequences, New Mexico; (3) Ile aux Aigrettes, Mauritius; (4) Grande Caverne, Rodrigues; (5) Pleistocene Park, northeastern Siberia, Russia; (6) Nelson Lakes National Park, New Zealand.

famous extinct denizen, the dodo (*Raphus cucullatus*), is a delightful place with beautiful beaches and friendly multilingual people of mixed ancestry. English is the official language, but through the vagaries of history and migration, people there mostly speak Creole among themselves, perfect French to French speakers, and of course perfect English. Most remarkably, perhaps, inhabitants of towns like Mahebourg are mostly people of Indian ancestry, and many of the rest are descendants of people brought, generally not of their own free will, from Madagascar and Africa. The official currency is the Mauritian rupee.

I borrowed a bicycle at the little hotel ("Sorry, sir, the boy's bicycle is engaged, but I do hope you won't mind this pink girl's model") and pedaled off along the coast to the little refuge's adjacent boat dock. A young Creole lady picked me up in the boat, along with a few tourists going out.

In the small museum on Ile aux Aigrettes are many paintings, including some excellent ones by my good friend the paleontologist and artist Dr. Julian Hume. Sadly, these are mostly images of creatures that were once there but are gone forever, like the dodo and the broad-billed parrot. Other displays were heartening, though, providing a photographic record of the reserve's activities on behalf of not just the few rare species that had survived there on the islet, but also many others such as the reintroduced Mauritian pink pigeon, one of their great success stories. These they reestablished through a project sponsored by the Jersey Wildlife Trust. They started with fewer than ten individuals in a surviving population on the mainland, encouraged them to produce dozens of offspring in captivity, then moved the fledglings to large fly cages on the reserve. When they are deemed ready, the cage doors are left open. For a while the big, handsome, intensely pink pigeons come back to sleep in the cage, and generally don't wander far. Eventually, our guide explained proudly, like a mother bragging about the accomplishments of her grownup children, they move to other parts of the islet and then on to the Mauritian mainland a few hundred meters across the water. Thanks to this "soft release" program this species of pigeon now numbers in the hundreds, and has been downlisted by the International Union for Conservation of Nature (IUCN) from "critically endangered" to just "endangered."

They have also implemented great conservation measures with rare Mauritian lizards. But what I really wanted to see (one of the projects I had been corresponding with these folks about for a couple of years by email) was the rare plant reforestation project, which was unforgettable in its simple cleverness. They had taken large boulders that were abundantly present nearby, and laid them out to make a minimally fenced area. What could you keep in with just a line of big boulders? An animal that was very important to the prehuman Mauritian ecosystems and got wiped out along with the dodo: giant tortoises. Nineteen of them were lumbering about inside the enclosure, eating the weedy understory, or just doing nothing (a favorite tortoise activity). But what happens to the poor plants, you are wondering? Each of the small native trees inside, wonderful little saplings of the rare and precious Mauritian ebony, tambalacoque, and other very rare trees and shrubs, had three old car tires encircling it, with rebar stakes driven through to hold them in place. Larger native trees, part of the site's legacy, were safe from the giants' beaks, because they don't chew bark. Everything else inside, a host of aggressive invading weeds, was being eaten to a nub, and even pulled up by the roots, by the great reptilian restorationists.

I said the giant tortoises of Mauritius were extinct. Well, not exactly. The populations on Mauritius were completely wiped out in the few short centuries of humans on Mauritius, as was also the case for the other Mascarene islands. But a very similar, if not the same, tortoise survives in great numbers on the uninhabited nature-reserve island of Aldabra north of Madagascar. DNA studies by Eric Palcovacs of Yale University have revealed that this living tortoise is likely the same species as one of the giant "extinct" species on Madagascar.[2] The theory goes that they evolved on Madagascar, then floated, like dome-shaped rafts with legs, to all these smaller islands, undoubtedly including Mauritius. On Mauritius, they had evolved to the "saddle-backed" form, with very thin light shells tipped up on the neck end to allow the hungry tortoise to reach up as high as possible into the trees.

Anybody with the necessary permits can have a few of these tortoises from Aldabra Sanctuary if they are willing to pay the shipping and handling, according to my guide. (It seems there are too many on Aldabra—well over 100,000.) They keep the vegetation there cropped pretty short, and many appear stunted from not getting enough to eat. What a difference being uninhabited by humans makes for an island, or any landmass . . . just what I have been trying to say throughout this book!

Meanwhile, this "tortoise ranching" approach to restoration in the Mascarene Islands has also thoroughly taken root in the private sector there. Owen Griffiths, a local zoologist and entrepreneur, has wonderful ecotourism projects at La Vanille on Mauritius and Grande Caverne on Rodrigues, the Mascarene island even farther east in the Indian Ocean. Visitors to these privately owned nature reserves see the rare plants and fruit bats of these remote islands, but the big attraction is the tortoises—not just dozens and dozens of adults, but hundreds and hundreds of babies at all stages, bred hatchery-style en masse, and viewed from a beautiful little trail. At La Vanille, the trail also leads through a small building where Griffiths keeps his absolutely spectacular and well-curated invertebrate collection, featuring not just the Mascarenes but their big neighbor to the west, Madagascar. At Grande Caverne, the François Leguat Giant Tortoise and Cave Reserve features a tour of a limestone cave system that contains fossils of the extinct creatures of Rodrigues, such as the solitaire (*Pezophaps solitaria*), a giant endemic flightless pigeon even bigger but less famous than the dodo of Mauritius. The sediments are full of bones of giant tortoises, and now giant tortoises roam the adjacent canyons again. The place is remarkably similar to Makauwahi Cave Reserve in concept. The idea emerged independently there—almost exactly halfway around the globe from Kaua'i, and in the opposite hemisphere.

These nice folks on Mauritius and Rodrigues can actually find much of the "prehuman" information needed to do a restoration just by looking in old books. European colonizers were eyewitnesses to this remarkable fauna, which initially lacked all fear of humans and was soon decimated. When the MWF and Owen Griffiths wanted to do something similar to what we are belatedly trying to do at Makauwahi Cave on Kaua'i, they didn't need to excavate and core (although they are very interested in this kind of research, which is why Mauritian researchers and conservationists contacted me in the first place, prompting me to stop here when in the neighborhood). All they had to do was read about it in books. Lucky them. There are only a handful of places like that in the entire world, mostly tiny remote oceanic islands and the polar regions. Everywhere else, this great initial human-induced transformation took place before the advent of writing and the practice of recording local natural histories and phenomena. Even in Mauritius, though, there is good reason to suspect that human-caused transformations were well along before things got settled there enough for people to start compiling a history, and much of the early history is generally problematic.[3]

As you can imagine, this work in Mauritius, New Zealand, and some other island nations is pretty uncontroversial most of the time. These are among conservation's global triumphs. Restoration work on these systems benefits from fairly recent reference points, and the plants and animals involved are considered national treasures. The controversy surrounding the idea of looking to the past to design future restorations is mostly a continental issue, one that has gotten mild-mannered paleoecologists into the midst of a big nasty fight.

It's time to define a word I have used only sparingly until now: *rewilding*. This is a term first popularized in North America by "deep ecology" thinkers like Dave Foreman, Reed Noss, and Michael Soulé. It is not so easy to define succinctly, but the general idea is that wild areas can preserve more biodiversity and ecological function by increasing their size and species richness, particularly focusing on keystone species that are essential to ecosystem functions. That doesn't do it justice, but here's where it gets controversial: in North America, this concept has so far taken the form of proposals to join large tracts of public, and in some cases private, lands under cooperative agreements, and proceed to reintroduce top carnivores such as wolves and charismatic megafauna such as bison, grizzly bears, and moose.[4]

This kind of large-scale restoration has become a minor battleground in the post-9/11 culture wars in the United States. Paleoecologists got mixed up in it through Paul Martin, of course. He has talked throughout his career about the unfortunate depauperation of the North American megafauna at the end of the Pleistocene. A really poetic piece by Martin, "The Last Entire Earth," broached the idea in 1992, drawing from Henry David Thoreau's musings about mastodons and past biodiversity losses.[5] "I listen to a concert in which so many parts are wanting," Thoreau said. "I take infinite pains to know all the phenomena of the spring, for instance, thinking that here I have the entire poem, and then, to my chagrin, I hear that it is but an imperfect copy that I possess and have read, that my ancestors have torn out many of the first leaves and grandest passages, and mutilated it in many places."

Paul and I published a paper in the same magazine, *Wild Earth* in 1999 that asked scientists and laymen to consider whether formerly North American creatures like horses, camels, and perhaps even elephants could have some sort of limited American future.[6] This idea has come to be known as Pleistocene rewilding, for better or for worse.

But the Russians got to the rewilding moon first. Sergei Zimov has been writing about restoring the unproductive moss tundra of the subarctic to productive grassland by restoring the megafauna since the early 1990s, and he is currently the director of a big piece of Siberia called—what else—Pleistocene Park.[7] Right now it's just moderately large fenced tracts with Yakutian horses, moose, American bison, musk oxen, and reindeer, but the plan calls for soon adding Siberian tigers and perhaps . . . who knows? Maybe someday, cold-adapted Indian elephants from the Himalayas and South African white rhinos (living relatives and ecological surrogates for the extinct mammoth and wooly rhinoceros, respectively). Some Japanese scientists are trying to fertilize Indian elephant eggs with frozen mammoth sperm from Siberia—trying, I said. Actually recovering usable complete genomes of DNA from fossils is exactly like trying to follow a roadmap after someone has put it through the shredder. But who knows, these are the Days of Miracle and Wonder. Anyway, the Japanese scientists think they may eventually succeed, yielding a half mammoth. Repeating this process with the eggs of the hybrid offspring, these scientists estimate, would produce a creature 88 percent mammoth in just 50 years. Dr. Zimov has made it clear that he is not involved with this scheme. He had to, I guess, because the Japanese scientists had already been quoted in the international media as having Zimov's Siberian restoration in mind as a place to release a reconstituted mammoth.

Now you see what I mean about how the logical or illogical extension, depending on point of view, of our Poor Man's Time Machine notions to the continents has the potential to change the biological world in some fundamental ways, or at least get some people really upset. The idea of rewilding in this "deeper" way, that is, transforming a landscape to something more like its past condition by bringing back the megafauna or its cousins or surrogates, is out now, and reproducing on its own. I organized a symposium in 2001 for the annual meeting of the Society for Conservation Biology, held in Hilo, Hawaii, called "Draining the Past to Irrigate the Future." I invited paleoecologists and land managers to submit papers on the topic of paleoecology guiding restorations. This was a big hit at the conference and led to an article that I wrote with two of the participants, Dave Steadman and Paul Martin, titled "Evolution's Second Chance."[8] Using examples from around the world, we argue for the idea of probing the fossil record as a guide, then proceeding to enrich the restored landscape with appropriate species. We had great examples from Dave's work in the South Pacific, and of course trotted out the Kaua'i results and talked about Paul's ideas for North America.[9] We suggested considering returning gray whales to the Atlantic (they've been absent only a couple of centuries) by towing them through the Panama Canal, and having a "Holocene Park" in Madagascar featuring not only the surviving giant crocodiles, but also African hippos, ostriches, and Aldabra tortoises, all of which previously had close relatives there.[10] We could use the tortoises to weed groves of native trees along the lines of the Mauritian model.

Tim Flannery echoed some of these ideas in his excellent popular treatment of North American ecology from the viewpoint of an Australian, *The Eternal Frontier.*[11] And media billionaire Ted Turner was starting to get interested. In September 2004, a dozen paleoecologists, conservation biologists, and large mammal ecologists were invited to one of Turner's legendary ranches to do some concentrated thinking about this big idea, to see how far it is possible and desirable to go in the direction of reconfiguring ecosystems that existed for millions of years but were snuffed out in North America about 13,000 years ago.

A graduate student at Cornell University, Josh Donlan, had been talking up the idea of a kind of think-tank exercise on Pleistocene rewilding for many months, and he had managed to put together some funding from the Turner Endangered Species Fund, the Environmental Leadership Program, and the Lichen Foundation, to invite these folks, including Paul Martin, Michael Soulé, Dave Foreman, Joel Berger, Harry Greene, Jane Bock, Carl Bock, Gary Roemer, Felisa Smith, and Jim Estes (in absentia). We ate, slept, and talked for three days in the hunting lodge at the Turner Ladder Ranch near Truth or Consequences, New Mexico. Although Ted wasn't there, several of his biologists were. I also quite accidentally had dinner the night before the conference with Ted's son Reed Beauregard (Beau) Turner, an avid outdoorsman like his father. He was there elk hunting; I showed up from Hawaii, as the farthest-traveling participant, the night before the conference to adjust to the time change. Naturally we found plenty to talk about. His dad owns more land than any other individual in the United States, including a swath of a dozen giant ranches from Montana to New Mexico. Beau is responsible for managing all this real estate. So what he thinks about this idea really could make a difference. The ranch where we stayed has no cattle anymore, but it does have more than 3,000 bison (eventually served up as bison burgers in Ted's restaurant chain, I guess; there are more than 30,000 bison on the Turner ranches combined), as well as free-roaming elk, burrowing owls, prairie-dog towns, and a wolf pack for reestablishment through a soft-release program not so different from the one for the Mauritian pink pigeon.

This informal meeting of "rewilders" was a great time, with lots of rich discussion. Opinions differed as to how far we could or should go with ideas like this, and how fast, and where, and how to pay for it. How could it be done without posing undue danger to the public, or having it perceived that way regardless of the realities? And then there's the contentious issue of how long ago is too long ago to count as an appropriate restoration target. In other words, restoration on Mauritius would be back four centuries, New Zealand eight, Hawaii ten, Madagascar twenty. But North America's restoration horizon, instead of being the five-century "Columbian Curtain" that Paul and I called European arrival in *Wild Earth,* would be 130 centuries. Maybe even for systems that develop over tens of millions of years, that is a significantly long time. We convinced ourselves that it was not, by many important evolutionary and ecological criteria, that long really. We also went out one night onto a remote ridge overlooking this vast ranch. Someone did a really good imitation of a wolf howl. From an enclosure the next ridge over, the real wolves howled back. Potentially reintroduced, *inter situ* wolves. Rewilded.

This group coauthored a commentary piece in the journal *Nature* summarizing some of our thoughts from the Turner conference.[12] Some of us, including me, worried that it might come off as too strident and maybe jostle the public psyche a little too hard. This was a lot of new information and radical proposals to process, even for scientists and conservationists. Anyway, when that thing came out, poor Josh Donlan and his adviser, herpetologist Harry Greene, found themselves at the center of a firestorm.

A week after publication, this article had generated more than 1,000 letters and phone calls from three continents and had been covered by the TV networks, most newspapers throughout the country, and a wide range of magazines. The response ranged from intrigued to laudatory to hateful, and the discussion in the media, conservation magazines, science journals, and on the Web has hardly trailed off. If you type into a major search engine "Pleistocene rewilding" the blizzard of blogs and media hits is astounding. One unfortunate coincidence was the initial appearance of this story in major news media the same day as reports of an American youngster being mauled to death by a captive tiger, so the two issues became related in the instantaneous media-consumer mind. Discussion turned pretty quickly from our modest emphasis on Bolson tortoises and wild horses to some of the animals our report pointed out would be the most controversial and also go the farthest, someday, after careful study and experimentation, in possibly reshaping ecosystems to something like past states. Elephants, lions, and even cheetahs were maligned in the press along with the crazy academics and wild-eyed deep ecologists who were now proposing to unleash elephants on Iowa cornfields or whatever. What had been couched in cautious terms of the need to think about ways we could experiment safely with megafaunal reintroductions was quickly overblown in the simplistic mass media and public mind as a proposal to give the country back to the . . . zoo animals? Some African conservationists reacted negatively from a whole different angle—these arrogant American scientists were going to transfer the African megafauna to their own affluent country and put the real African game parks and the safari ecotourism industry there out of business.

These were criticisms we had worried a lot about at the Turner Ladder Ranch, but we decided to move ahead cautiously. Some of us weren't so sure anymore after the verbal drubbing rewilding had received in such normally idea-friendly venues as National Public Radio's *Talk of the Nation: Science*

Friday. Josh was new at this game, and perhaps didn't take Paul and me seriously enough when we recounted at the ranch the storm of discussion that our 1999 *Wild Earth* article, and then Tim Flannery's book *Eternal Frontier*, had already generated in the media, followed by the usual die-down after a few weeks. At the meeting, I fretted that we might do well to emphasize the mild-mannered island examples of rewilding a little more, and try to play down the idea of big scary African megafauna roaming the West, even on private ranches with state-of-the-art mega-fences. The Turner folks there certainly preferred this ultra-cautious approach. Ted didn't need us getting him into trouble with everyone from his ranch neighbors to the international media over rumors that he might even vaguely be considering trying an elephant or two, or starting a cheetah population to control deer herds on his property. The Bolson tortoise might be possible someday, but even that needed to be done with all proper deliberation.

Some of us also predicted two other ways our ideas would get twisted up. First, and most ironically, those of us who have spent so much of our lives documenting and battling the effects of invasions by alien species, were now in a sense defending some. The horse and its relatives evolved in North America for 40 million years, only to be displaced by end-Pleistocene, probably human-mediated, extinction. Whether it is allowing wild ponies on the Outer Banks of North Carolina or controlling burros in the Grand Canyon, or formulating policy for mustangs in western national parks, this view of equids as not American invaders but returning natives is a divisive issue within the conservation and land management community.

The second dark prophecy from our little spontaneous think tank also has unfortunately become true. Talk of yet more big animals roaming the West on larger tracts of the range landed our little time machine in a place we had no interest in being: in the negative spotlight of what has come to be known as the property-rights movement. What started in our minds as the well-meaning technical question of how past evidence might yield ideas for a better future for ecosystems in need of repair was predictably seen by some on the far right as a new twist in the Green conspiracy to take over much of the country for wacky purposes only radical environmentalists could fathom. Elephants might be the Republican mascot, but they were not meant to be set loose in the backyards of the property-rights mentality of Middle America. I can't resist quoting a bit from one of the blogs on this, titled "The Pleistocene Park Project—Removing Civilization from North America," by Robert J. Smith, adjunct environmental scholar for the Competitive Enterprise Institute. This was a speech from the Ninth Annual National Conference of Private Property Rights, sponsored by the Property Rights Foundation of America, Inc.:[13]

> Well, on Thursday, August 18 . . . the world's most prestigious science journal, called *Nature,* hit the news stands. It contained a remarkable article written by twelve authors, ecologists and conservationists . . . and it called for the creation of something called the U.S. Ecological History Park . . . This idea has come from a joint vision of a number of groups. The radical Greens and so-called monkey wrenchers, if you know who they are. Also effete urban planners at some of the eastern universities, the deep ecologists who sit around and figure out what's the relationship of man to the rest of the world, and the saving-the-creation religionists. While the article surely jolted the world's media for about a week this summer, there was no cry of outrage from the scientific community at all. And it quickly became nicknamed the Pleistocene Park. Leading conservationists and ecologists called it "a bold vision." It's certainly that. It "will fill long vacant niches." It will "return U.S. grasslands to their evolutionary health." And we are trying "to look at a bigger conservation pattern." And they also say, well, it will bring in eco-tourism.
>
> Well, it may bring some eco-tourists, perhaps, but it will also make it rather dangerous to step outside your door early in the morning and let your kids go outside if a lion decides to eat your kids. We already had a couple of problems with that with grizzly bears and the Endangered Species Act in Montana.

In the minds of some, our time machine has already run amok. The Pleistocene is climbing over the fence, or scratching under, and like those grizzlies and the Endangered Species Act, looking to eat your kids. I prefer the summarization of what we are advocating that appeared in an article by William Stolzenburg in the respected journal *Conservation in Practice:* "The Ladder group agreed on several sobering premises: That human influence had utterly pervaded the planet. That what qualifies for wildness today is a paltry façade of the awesome Pleistocene bestiary we stumbled upon only 13,000 years ago. That the difference between then and now is at least partly, if not principally, our own doing and therefore our duty to repair."[14]

All we are saying, I guess, is give evolution a chance. But less than half the American public, most studies indicate, even believes in evolution, so why try to restore it? On the other hand, don't forget that Thomas Jefferson, a kind of creationist in the technical sense at least, on meeting the mastodon's remains, was sure God would never let such a fine creature become extinct.[15] My mother has always been fond of saying, "God works in mysterious ways . . ." I was never quite sure what she meant by that, but I guess that was part of the mystery. Paul Martin uses the term *resurrection ecology* in his new book *Twilight of the Mammoths.* And there really is a "Save-the-Creation" movement. So the Poor Man's Time Machine plows on through stormy uncharted waters, even vaguely theological . . . while Paul Martin and the gang of twelve "Ladderites" try to remove civilization from North America . . .

Of all the thousands of people to tour Makauwahi Cave over the 17 years we have been working there, perhaps the most exciting visit for Lida and me was not some politician or popular celebrity, but Paul Martin, of course. The mere fact of the man coming there who has played such a big role in the discussions of a half century regarding how things change when humans arrive—that is not remarkable. Paul has visited sites all over the world that contribute to his story about human transformation of the biotic landscape. He even scouted out some of my key fossil sites in Madagascar 20 years before I got there. But there is a feature about Paul that perhaps you would not have guessed.

The guy leading the charge can hardly walk. He contracted polio while doing graduate work in Mexico in the 1950s, and has had scant use of his legs since. Our cave is not so remote for a walker, but it is perhaps beyond the limit of most folks who get around on two canes and sometimes a wheelchair. "Most people crawl *into* caves, Dave," he remarked with a grin when refusing a balloon-tire wheelchair Lida had borrowed from a lifeguard friend of ours, "but I also can crawl *to them*." So, in the summer of 2001, after the "Irrigating the Future" symposium in Hilo, Paul and his wife, Mary Kay O'Rourke, also a palynologist but specializing in allergy-related pollen studies, came to visit us on Kaua'i. After a crawling tour of the cave, Paul sat in a nice canvas folding chair Mary Kay brought in and watched me dig all day in the East Pit. I have spent many nice days with Paul before and since then, but that one was a fine memory. Three years later at the Turner Ladder Ranch, Paul could talk from firsthand experience about his rewilding dream and include Kaua'i and other small remote islands as examples of his grand plan for restarting evolution—writ smaller, but already quite far along compared with the continents. And the deluxe canvas chair he left behind we have made good use of down there ever since. We call it the Paul Martin Endowed Chair.

Paul is one of my favorite people, obviously. Statuesque, soft-spoken, self-effacing, Paul has gently pressed his strong scientific opinions in the face of constant controversy, and generously supported the careers of followers and critics alike. "Science doesn't move forward unless someone is willing to paint a target on himself, Dave," Paul explained one time when I asked how it felt to be at the center of so much scientific argumentation. Of course, he learned from the best. After earning his Ph.D. in zoology at the University of Michigan in 1955, he completed a postdoctoral fellowship at Yale with Ed Deevey, one of the early giants of ecology and the scientist who introduced paleoecology as a science in America. (Professor Deevey was also my adviser Dan Livingstone's Ph.D. adviser and a G. Evelyn Hutchinson student.) Paul has been working at the Desert Laboratory of the University of Arizona ever since, as research associate, assistant professor, associate professor, professor, and, since 1989, professor emeritus in the Department of Geosciences. His work has spanned arid-lands palynology, geochronology, large-mammal paleoecology, . . . "and large animal dung," he would add with a good-natured laugh on himself. His books, journal articles, and edited volumes on Quaternary extinctions are the industry standard, so to speak. And I believe he owns the world's largest reference collection of large animal dung, both fossil and modern.

One could probably sum up the current situation for the Makauwahi Cave Reserve and the National Tropical Botanical Garden by saying that we have followed Paul's advice and painted some targets on ourselves, individually and as institutions. We are not proposing anything in the Hawaiian Islands that is at all likely to eat your kids, but there is a controversial assertion being made that needs to be stated explicitly: *A concentrated effort by humans can slow a human-caused extinction catastrophe and recover some lost ecological function and biodiversity.*

On the surface, like Pleistocene rewilding, this seems an innocuous, academic notion. But to actually take on such a task requires a level of optimism that may be unjustified and a certain technical expertise that simply may not be available. Above all, perhaps, stopping an extinction event can be expensive financially and politically. Restarting evolution, or rescuing Creation (take your pick) is probably labor intensive. Remember, too, anything involving rare plants and animals is now intensely political in America—many people worry about having the Endangered Species Act roaming their yard, both literally and figuratively.

Many conservation organizations and government agencies are at work saving Hawaiian species, of course. Their techniques run the gamut from the University of Hawaii's many basic and applied research programs to The Nature Conservancy's preservation and management of relatively pristine reserves in the remote highlands, to the Bernice P. Bishop Museum's painstaking documentation of biotic losses and the U.S. Department of Agriculture's Natural Resources Conservation Service, subsidizing farmers who grow native plants (like Lida). The National Tropical Botanical Garden, however, has an impressively comprehensive strategy to save Hawaii's rarest native plants through a complex horticultural protocol that sounds an awful lot like—you guessed it—*rewilding* Hawaiian landscapes. True,

NTBG manages relatively pristine reserves, such as more than 400 acres (162 ha) in Upper Limahuli Valley, an uninhabited and virtually unreachable area (without a helicopter), perched above an 800-foot (244 m) waterfall and hemmed in by Kaua'i's highest ridges and the vast Alaka'i Swamp. But NTBG's Conservation Department has embarked on a program aimed at pulling Hawaii's rarest plant species, plants represented by fewer than 50 individuals in the wild, back from the brink of extinction.

It's a multistep process, requiring integration of a staff of at least two dozen people whose jobs are each somewhere along the continuum from finding the rarest plants and collecting their seeds to propagating these plants in state-of-the-art greenhouse and micropropagation (tissue culture) facilities, to keeping up with their pedigree, whereabouts, and fate. We plant our rare babies in living (*ex situ*) collections in the botanical gardens, and also turn them out into the "wild." We use them to enrich existing native populations, but what's more radical is that we are quite deliberately *creating new populations of rare species.* We do so by taking lands invaded by nonnative plants and replacing, through a variety of methods, the weed species with common natives and ultimately, as the site stabilizes, with incipient populations of rare plants created from the largest appropriate gene pool of parents possible. Just like the Makauwahi Cave Reserve.

What is so exciting about this for me, other than standing in the bull's-eye of this particular target, is that being the NTBG director of conservation means an incredible array of talent in field botany, plant propagation, and ecological restoration is available to me in the form of this fantastic staff of "Extinction Busters." If anybody can do it, it's these folks. It also counts for a lot that NTBG has a powerful base of support in terms of talent, money, know-how, and influence from its remarkably large and diverse assemblage of trustees and fellows. Add to this NTBG's exemplary volunteer program, Na Lima Kokua ("The Helping Hands") and the sage guidance of CEO Chipper Wichman and a crack team of financial managers, scientists, and educators, and it's almost as if our Poor Man's Time Machine has docked at a conservation Mother Ship. NTBG is an organization with the clout needed to stop a mass extinction, if one can be stopped.

There are some criticisms around worth considering. Groups like NTBG are committed to pursuing what we have been referring to as *inter situ* conservation—that is, in-ground restorations outside the current or recent historical range of a species, but within the prehistoric range. This puts conservation in some tricky gray areas. Where do you draw the line chronologically? How far back is too far back to use as justification? Can you ever justify reestablishing a species or close relative on a landmass that it has not been recorded from historically, but only as a recent fossil? That is what we have done with the extremely rare native palm *Pritchardia aylmer-robinsonii* from the island of Niʻihau, Kauaʻi's neighbor. Does this kind of "farming" of endangered species raise false hopes in the conservation-conscious public, diminishing enthusiasm for in-situ conservation and preservation? If techniques are devised that really allow us to fix a damaged ecosystem, doesn't that just encourage developers to be incautious, since errors are correctable? Is this a palliative strategy that lulls the public into believing no further sacrifice is required to save biodiversity, since extinction is virtually, or even actually in some cases, reversible? I think we have good enough answers for these questions, based on experience and the best science available, to keep moving forward, and we will know considerably more about how best to rewild Hawaii in just a few short years.

On a more technical level, I still worry a lot about invasions. Is it possible that biological invasion is such a pervasive and ever escalating problem on isolated islands like Kauaʻi that nothing can really be done about it? Is restoration possible in the face of constant new arrivals? Can the incidence of new arrivals be drastically slowed through agricultural inspection at the ports of entry? Can *any* native species hold out in the long run?

What about "environmental surprises"? Can we anticipate and absorb the effects of future large-scale events, such as global warming, sea-level rise, hurricanes as severe as ʻIniki or worse, the once in a half-millennium tsunamis, flood and drought, and so on? What will be the effect of the economic and energy-based catastrophes that seem to be looming?

These are tough, important questions. So far, we know this much: rewilding even one small part of one small island is hard work for many willing hands, and it could take a long time or in a sense never be finished. Makauwahi Cave has taught us that. It has also taught us, beyond all doubt, that trying to make a good place better, however small and incomplete the effort, can be very deeply gratifying on the personal level. One cannot help musing, in the midst of this pleasant labor, whether humans could do this on a much bigger scale. Perhaps we will never know. Perhaps rewilding or even moderate large-scale restoration at regional or continental scale is an illusion, or an expensive toy for the rich. And what about sustainable development? Is that possible at any scale, or is it just a comforting myth or, worse still, a dangerous delusion? I don't think so, but a lot of people do.

Successful human colonizers come ashore to an uninhabited land only once, and thereafter the place has changed forever, at least on our time scale. So human arrival is a very significant event for nature. I am nearly certain, too, that humans at any technological level from the Stone Age onward have the capacity to drastically alter any ecosystem, no matter how large, remote, or toothsome, in a few generations. They have had this capacity for tens of thousands of years and maybe even longer. And, even in the face of a strong, positive tradition aimed at preserving the environment, like that of the ancient Hawaiians, people change the natural systems they inhabit anyway. Something about human cognition is destabilizing for nature in places that evolved without humans, apparently. Some other species may modify their environment for their own benefit, but the world has never seen anything like us.

It remains to be seen whether the pattern I have described in this book, this deadly syncopation—humans arrive, nature diminishes—has a degree of symmetry. By that I mean, wouldn't it be a big relief if we could discover, over the next few critical years, that humans can, in a similar stretch of a few generations, actually stop wiping out the rest of nature while maintaining or even improving the human standard of living? We will not know the answers to these questions, most likely, until it is too late to do anything about the outcome. If we don't try as hard as we can in the meantime to get the answers and apply them, it is certainly already too late. We humans can really pack a wallop, and have, time and time again, across the face of the planet.

Glossary

Diatoms: Unicellular or colonial algae with a cell wall reinforced with a skeleton of silica that easily fossilizes.

Eolianite: A type of cemented sandstone or calcarenite that is formed from sands accumulated in wind-driven dune formations.

Holocene: The second and present epoch of the Quaternary period, following the last glaciation and extending from roughly 11,000 years ago to the present.

Megafauna: Large animals, sometimes defined as those having a body weight of 100 pounds (44 kg) or greater.

Paleoecology: The study of past environments using evidence from sediments, fossils, artifacts, oral traditions, and historical documents.

Pleistocene: The first epoch of the Quaternary period, spanning the time from 1.8 million years ago to about 11,000 years ago. This geological term roughly corresponds to the colloquial "Ice Ages."

Quaternary: The present and most recent period of the Cenozoic Era. It encompasses the Pleistocene and Holocene, and extends from 1.8 million years ago to the present.

Radiocarbon years vs. calendar years: Radiocarbon dates are given in the text as a number plus or minus its statistical standard deviation, as measured in the lab with the current standard corrections. Actual calendar years in the text may be given as a range, which is the 95 percent confidence interval for the radiocarbon date with the current standard calibration model applied. For details see the original publications cited with the dates.

Stratigraphy: The study of layers in rocks or sediments.

Subphreatic: In or under the water table. Well water is phreatic water.

Notes

Chapter 1. Time's Most Important Moment

1. Martin and Steadman, "Prehistoric Extinctions," 17; Burney and Flannery, "Fifty Millennia of Catastrophic Extinctions," 395.
2. Wagner et al., "Hawaiian Vascular Plants at Risk," 64.
3. Burney, "Mysterious Island," 50.
4. Martin, "Africa and Pleistocene Overkill," 339; Martin and Wright, eds., *Pleistocene Extinctions,* 75; Mosimann and Martin, "Simulating Overkill by Paleoindians," 304.
5. Livingstone, "Lightweight Piston Sampler for Lake Deposits," 137.
6. Burney, *Late Quaternary Environmental Dynamics of Madagascar;* Burney et al., "Chronology for Late Prehistoric Madagascar," 25.

Chapter 2. Proverbial Tracks

1. Burney et al., "Holocene Charcoal Stratigraphy of Puerto Rico," 273; Burney et al., "Holocene Record from Maui," 209.
2. Olson and James, "Fossil Birds from the Hawaiian Islands," 633; Olson and James, *Description of 32 New Species of Birds;* Olson and James, "Prehistoric Status and Distribution of the Hawaiian Hawk," 65; James and Olson, "*Description of 32 Species of Birds, II;* James and Burney, "Diet and Ecology of Hawaii's Extinct Flightless Waterfowl," 279; James et al., "Radiocarbon Dates on Bones of Extinct Birds from Hawaii," 2350; James, "Osteology and Phylogeny of the Hawaiian Finch Radiation," 207.
3. Tyson, *The Eighth Continent.*
4. Arthur, *Tales of the Tempest.*

Chapter 3. Constructing a "Poor Man's Time Machine"

1. Burney, "Tropical Islands as Paleoecological Laboratories," 437.
2. Burney et al., "Chronology for Late Prehistoric Madagascar," 25.
3. Bliley and Burney, "Late Pleistocene Climatic Factors in the Genesis of a Carolina Bay," 83.
4. Semonin, *American Monster.*

Chapter 5. Opening Ancient Doors

1. Burney et al., "Holocene Assemblage from Kaua'i," 615; James and Olson, "The Diversity and Biogeography of Koa-finches," 527; James and Olson, "A New Species of Hawaiian Finch," 335.
2. Leopold, *Sand County Almanac;* Martin, "Thinking Like a Canyon," 82.
3. Blay and Siemers, *Kauai's Geologic History.*
4. Hearty et al., "Stratigraphy and Whole-rock Amino Acid Geochronology," 423; Blay and Longman, "Stratigraphy and Sedimentology," 93.
5. Juvik and Juvik, *Atlas of Hawaii.*
6. Brook et al., "Desert Paleoenvironmental Data from Cave Speleothems," 311; Burney et al., "Holocene Pollen Record for the Kalahari Desert," 225; Burney et al., "Environmental Change, Extinction, and Human Activity," 755.
7. Burney et al., "Holocene Assemblage from Kauai," 615.
8. Burney and Kikuchi, "Millennium of Human Activity," 219.

Chapter 6. Characters and a Stage, but No Script

1. Burney, "Theories and Facts Regarding Holocene Environmental Change," 75; Burney, "Madagascar's Prehistoric Ecosystems," 47; Burney et al., "Chronology for Late Prehistoric Madagascar," 25; Robinson et al., "Landscape Paleoecology and Megafaunal Extinction," 295.
2. Dahl, *Malgache et Maanyan;* Diamond, *Guns, Germs, and Steel;* McGovern, "Cows, Harp Seals, and Churchbells," 245.
3. Burney, "Rates, Patterns, and Processes," 145.
4. Davis, "Spores of the Dung Fungus *Sporormiella,*" 290.
5. Burney et al., "*Sporormiella* and the Late Holocene Extinctions," 10800.
6. Soulé, *Conservation Biology.*
7. Robinson et al., "Landscape Paleoecology and Megafaunal Extinction," 295.
8. Kikuchi, "Prehistoric Hawaiian Fishponds," 295.

Chapter 7. Fishponds

1. Burney, "Twelve Sites on Kauai," 13.
2. Ching et al., "Archaeology of Puna," 123.
3. Burney, "Twelve Sites on Kauai."
4. Burney and Burney, "Charcoal Stratigraphies for Kauai," 211.
5. Kirch, *Feathered Gods and Fishhooks.*
6. Masse and Tuggle, "The Date of Hawaiian Colonization," 229; Athens et al., "Avifaunal Extinctions, Vegetation Change, and Polynesian Impacts," 57.

Chapter 8. A Snail's Tale

1. Burney et al., "Holocene Assemblage from Kaua'i," 615.

Chapter 9. Mauka Marshes

1. Hutchinson, *The Ecological Theater.*
2. Burney et al., "Holocene Record from Maui," 209.
3. Selling, *Hawaiian Spores;* Selling, *Hawaiian Pollen;* Selling, *Vegetation History;* Athens et al., "Holocene Lowland, Oahu," 9; Athens et al., "Prehistoric Hawaii," 57; Athens, "Hawaiian Lowland Vegetation," 248; Hotchkiss, *Quaternary of Hawaii;* Hotchkiss and Juvik, "Pollen Record from Oahu," 115.
4. Burney, "Twelve Sites on Kaua'i," 13; Burney and Burney, "Charcoal Stratigraphies," 211.
5. Burney, "Twelve Sites on Kaua'i"; Burney and Burney, "Charcoal Stratigraphies."
6. Juvik and Juvik, *Atlas of Hawaii,* 3rd ed.

Chapter 10. So What Happened, Anyway?

1. Olson and James, "Fossil Birds from Hawaii," 633; Burney et al., "Holocene Assemblage from Kaua'i," 615.
2. MacPhee, ed., *Extinctions in Near Time.*
3. Quammen, *Song of the Dodo.*
4. Martin, "Prehistoric Overkill," 354.
5. Cook, *The Journals;* Vancouver, *Discovery.*
6. Stannard, *Before the Horror.*
7. Vitousek et al., "Soils, Agriculture, and Society," 1665; Burney, "Twelve Sites on Kaua'i," 13.
8. Pratt, et al., *Birds of Hawaii.*

Chapter 11. Greetings from Old Kaua'i

1. Kikuchi et al., "Archaeological Excavations"; Burney and Kikuchi, "Millennium of Human Activity," 219.
2. Malo, *Hawaiian Antiquities;* Kalakaua, *Legends and Myths.*
3. Masse and Tuggle, "Date of Hawaiian Colonization," 229.
4. Bennett, *Archaeology of Kaua'i.*
5. Flannery, *Future Eaters.*
6. Burney and Kikuchi, "Millennium of Human Activity," 219.
7. Ibid.

8. Lahainaluna Papers (1885), manuscript #17, B. P. Bishop Museum Archives, Honolulu.
9. Tanimoto, *Return to Mahaulepu.*

Chapter 12. Irrigating the Future

1. National Tropical Botanical Garden, *Strategic Directions for the 21st Century.*

Chapter 14. Right Here, Right Now

1. Olson and James, "Fossil Birds," 633; Olson and James, "Role of Polynesians," 768; Olson and James, *Descriptions of 32 New Species;* James and Burney, "Diet and Ecology," 279.
2. Burney and Burney, "Paleoecology and 'Inter-situ' Restoration," 483.
3. Duffy and Kraus, "Science in Hawaii's Extinction Crisis," 3; Burney et al., "Evolution's Second Chance," 12; Blixt, "Conservation Methods"; Guerrant et al., *Ex-situ Plant Conservation.*
4. Burney and Burney, "Paleoecology and 'Inter-situ' Restoration," 483.
5. Soulé, *Conservation Biology.*
6. Caro, "Pleistocene Re-wilding Gambit," 281.
7. Burney et al., "Holocene Assemblage from Kaua'i," 615.
8. Burney and Kikuchi, "Millennium of Human Activity," 219.
9. Burney et al., "Holocene Assemblage from Kaua'i," 615; Erickson and Puttock, *Wetland Field Guide,* 51.

Chapter 15. Finding a Future in the Past

1. Atkinson, "Introduced Mammals and Models for Restoration," 81.
2. Palcovacs, "Evolutionary Origin of Indian Ocean Tortoises," 216.
3. Hume, "History of the Dodo," 65.
4. Foreman, *Rewilding North America.*
5. Martin, "Last Entire Earth," 29.
6. Martin and Burney, "Bring Back the Elephants!" 57.
7. Zimov et al., "Steppe-tundra Transition," 765; Zimov, "Pleistocene Park," 796.
8. Burney et al., "Evolution's Second Chance," 12.
9. Steadman, "Prehistoric Extinctions," 1123; Steadman and Martin, "Late Quaternary Extinction," 133; Steadman, *Extinction and Biogeography.*
10. Burney, "Madagascar's Prehistoric Ecosystems," 47.
11. Flannery, *Eternal Frontier.*
12. Donlan et al., "Rewilding North America," 913.

13. Smith, "Pleistocene Park Project."
14. Stolzenburg, "Where the Wild Things Were," 28.
15. Semonin, *American Monster.*

Bibliography

Arthur, S. H. *Tales of the Tempests: The Hurricanes of Kaua'i.* Lihue: Primitive Graffiti Publications, 2001.

Athens, J. S. "Hawaiian native lowland vegetation in prehistory." In *Historical Ecology in the Pacific Islands,* ed. P. V. Kirch and T. L. Hunt, pp. 248–270. New Haven: Yale University Press, 1997.

Athens, J. S., H. D. Tuggle, J. V. Ward, and D. J. Welch. "Avifaunal extinctions, vegetation change, and Polynesian impacts in prehistoric Hawai'i." *Archaeology in Oceania* 37 (2002): 57–78.

Athens, J. S., J. Ward, and S. Wickler. "Late Holocene lowland vegetation, O'ahu, Hawaii." *New Zealand Journal of Archaeology* 14 (1992): 9–34.

Atkinson, I. A. E. "Introduced mammals and models for restoration." *Biological Conservation* 99 (2001): 81–96.

Bennett, W. C. *Archaeology of Kauai.* Honolulu: B. P. Bishop Museum Bulletin 80, 1931.

Blay, C., and R. Siemers. *Kauai's Geologic History: A Simplified Guide.* Kauai: TEOK Investigations, 1998.

Blay, C. T., and M. W. Longman. "Stratigraphy and sedimentology of Pleistocene and Holocene carbonate eolianites, Kaua'i, Hawai'i, USA." SEPM Spec. Pub. 71 (2001): 93–115.

Bliley, D. J., and D. A. Burney. "Late Pleistocene climatic factors in the genesis of a Carolina Bay." *Southeastern Geology* 29 (1988): 83–101.

Blixt, S. "Conservation methods and potential utilization of plant genetic resources in nature conservation." In *Integration of Conservation Strategies of Plant Genetic Resources in Europe,* ed. F. Begemann and K. Hammer. Gatersleben, Germany: IPK and ADI, 1994.

Brook, G. A., D. A. Burney, and J. B. Cowart. "Desert paleoenvironmental data from cave speleothems with examples from the Chihuahuan, Somali-Chalbi, and Kalahari deserts." *Palaeogeography, Palaeoclimatology, Palaeoecology* 76 (1990): 311–329.

Brook, G. A., D. A. Burney, and J. B. Cowart. "Paleoenvironmental data for Ituri, Zaire, from sediments in Matupi Cave, Mt. Hoyo." *Virginia Museum of Natural History Memoirs* 1 (1990): 49–70.

Burney, D. A. "Life on the cheetah circuit." *Natural History* 91 (1982): 50–59.

Burney, D. A. *Late Quaternary Environmental Dynamics of Madagascar.* Ph.D. dissertation, Duke University. Pub. No. 87-06828. Ann Arbor, Mich.: University Microfilms International, 1986.

Burney, D. A. "Theories and facts regarding Holocene environmental change before and after human colonization." In *Natural Change and Human Impact in Madagascar,* ed. S. M. Goodman and B. D. Patterson, pp. 75–89. Washington, D.C.: Smithsonian Press, 1997.

Burney, D. A. "Tropical islands as paleoecological laboratories: Gauging the consequences of human arrival." *Human Ecology* 25 (1997): 437–457.

Burney, D. A. "Rates, patterns, and processes of landscape transformation and extinction in Madagascar." In *Extinctions in Near Time: Causes, Contexts, and Consequences,* ed. R. MacPhee, pp. 145–164. New York: Plenum/Kluwer, 1999.

Burney, D. A. "Late Quaternary chronology and stratigraphy of twelve sites on Kaua'i." *Radiocarbon* 44 (2002): 13–44.

Burney, D. A. "Madagascar's prehistoric ecosystems." In *The Natural History of Madagascar,* ed. S. Goodman and J. Benstead, pp. 47–51. Chicago: University of Chicago Press, 2003.

Burney, D. A., G. A. Brook, and J. B. Cowart. "A Holocene pollen record for the Kalahari Desert of Botswana from a U-series dated speleothem." *The Holocene* 4 (1994): 225–232.

Burney, D. A., and L. P. Burney. "Paleoecology and 'inter situ' restoration on Kaua'i, Hawai'i." *Frontiers in Ecology and the Environment* 5 (2007): 483–490.

Burney, D. A., L. P. Burney, L. R. Godfrey, W. L. Jungers, S. M. Goodman, H. T. Wright, and A. J. T. Jull. "A chronology for late prehistoric Madagascar." *Journal of Human Evolution* 47 (2004): 25–63.

Burney, D. A., L. P. Burney, and R. D. E. MacPhee. "Holocene charcoal stratigraphy from Laguna Tortuguero, Puerto Rico, and the timing of human arrival on the island." *Journal of Archaeological Science* 21 (1994): 273–281.

Burney, D. A., R. V. DeCandido, L. P. Burney, F. N. Kostel-Hughes, T. W. Stafford, Jr., and H. F. James. "A Holocene record of climate change, fire ecology, and human activity from montane Flat Top Bog, Maui." *Journal of Paleolimnology* 13 (1995): 209–217.

Burney, D. A., and T. F. Flannery. "Fifty millennia of catastrophic extinctions after human contact." *Trends in Ecology and Evolution* 20 (2005): 395–401.

Burney, D. A., H. F. James, L. P. Burney, S. L. Olson, W. Kikuchi, W. L. Wagner, M. Burney, D. McCloskey, D. Kikuchi, F. V. Grady, R. Gage, and R. Nishek.

"Fossil evidence for a diverse biota from Kaua'i and its transformation since human arrival." *Ecological Monographs* 71 (2001): 615–641.

Burney, D. A., H. F. James, F. V. Grady, J.-G. Rafamantanantsoa, Ramilisonina, H. T. Wright, and J. B. Cowart. "Environmental change, extinction, and human activity: evidence from caves in NW Madagascar." *Journal of Biogeography* 24 (1997): 755–767.

Burney, D. A., and W. K. P. Kikuchi. "A millennium of human activity at Makauwahi Cave, Maha'ulepu, Kaua'i." *Human Ecology* 34 (2006): 219–247.

Burney, D. A., G. S. Robinson, and L. P. Burney. "*Sporormiella* and the late Holocene extinctions in Madagascar." *Proceedings of the National Academy of Sciences, USA* 100 (2003): 10800–10805.

Burney, D. A., D. W. Steadman, and P. S. Martin. "Evolution's second chance: Forward-thinking paleoecologists advocate jump-starting diminishing biodiversity." *Wild Earth* 12 (2002): 12–15.

Burney, L. P., and D. A. Burney. "Charcoal stratigraphies for Kaua'i and the timing of human arrival." *Pacific Science* 57 (2003): 211–226.

Caro, T. "The Pleistocene rewilding gambit." *Trends in Ecology and Evolution* 22 (2006): 281–283.

Ching, F. K. W., P. B. Griffin, W. K. Kikuchi, W. A. Albrecht, J. D. Belshé, and C. Stauder. *The Archaeology of Puna, Kaua'i.* Honolulu: Archaeological Research Center Hawaii, 1973.

Cook, J. *James Cook: The Journals,* ed. Philip Edwards. Prepared from the original manuscripts by J. C. Beaglehole, 1955–1967. London: Penguin Books, 2003.

Dahl, O. *Malgache et Maanyan.* Oslo: Egede Instituttet, 1951.

Davis, O. K. "Spores of the dung fungus *Sporormiella:* Increased abundance in historic sediments and before Pleistocene megafaunal extinction." *Quaternary Research* 28 (1987): 290–294.

Diamond, J. *Guns, Germs, and Steel: The Fates of Human Societies.* New York: W. W. Norton, 1997.

Donlan, C. J., J. Berger, C. E. Bock, J. H. Bock, D. A. Burney, J. A. Estes, D. Foreman, P. S. Martin, G. W. Roemer, F. A. Smith, M. E. Soulé, and H. W. Greene. "Pleistocene rewilding: An optimistic vision for 21st century conservation." *American Naturalist* 168 (2006): 660–681.

Donlan, C. J., H. W. Greene, J. Berger, C. E. Bock, J. H. Bock, D. A. Burney, J. A. Estes, D. Foreman, P. S. Martin, G. W. Roemer, F. A. Smith, and M. E. Soulé. "Rewilding North America." *Nature* 436 (2005): 913–914.

Donlan, C. J., and P. S. Martin. "Role of ecological history in invasive species management and conservation." *Conservation Biology* 18 (2004): 267–269.

Duffy, D. C., and F. Kraus. "Science and the art of the solvable in Hawaii's extinction crisis." *Environment Hawaii* 16 (2006): 3–6.

Erickson, T. A., and C. F. Puttock. *Hawaii Wetland Field Guide.* Honolulu: Bess Press, 2006.

Flannery, T. F. *The Future Eaters: An Ecological History of the Australasian Lands and Peoples.* New York: George Braziller, 1995.

Flannery, T. F. *The Eternal Frontier: An Ecological History of North America.* New York: Atlantic Monthly Press, 2001.

Foreman, D. *Rewilding North America: A Vision for Conservation in the 21st Century.* Washington, D.C.: Island Press, 2004.

Guerrant, E. O., K. Havens, and M. Maunder. *Ex-situ Plant Conservation: Supporting Survival in the Wild.* Washington, D.C.: Island Press, 2004.

Hearty, P. J., D. S. Kaufman, S. L. Olson, and H. F. James. "Stratigraphy and whole-rock amino acid geochronology of key Holocene and last interglacial carbonate deposits in the Hawaiian Islands." *Pacific Science* 54 (2000): 423–442.

Hotchkiss, S. C. *Quaternary Vegetation and Climate of Hawai'i.* Ph.D. dissertation. St. Paul, Minnesota: University of Minnesota, 1998.

Hotchkiss, S. C., and J. O. Juvik. "A Late-Quaternary pollen record from Ka'au Crater, Oahu, Hawaii." *Quaternary Research* 52 (1999): 115–128.

Howarth, F. G. "The cavernicolous fauna of the Hawaiian lava tubes. 1. Introduction." *Pacific Insects* 15 (1973): 139–151.

Hume, J. P. "The history of the Dodo (*Raphus cucullatus*) and the penguin of Mauritius." *Historical Biology* 18 (2006): 65–89.

Hutchinson, G. E. *The Ecological Theater and the Evolutionary Play.* New Haven: Yale University Press, 1965.

James, H. F. "The osteology and phylogeny of the Hawaiian finch radiation (Fringillidae: Drepanidini), including extinct taxa." *Zoological Journal of the Linnean Society* 141 (2004): 207–255.

James, H. F., and D. A. Burney. "The diet and ecology of Hawaii's extinct flightless waterfowl: Evidence from coprolites." *Biological Journal of the Linnean Society* 62 (1997): 279–297.

James, H. F., and S. L. Olson. "Descriptions of thirty-two new species of birds from the Hawaiian Islands: Part II. Passeriformes." *Ornithological Monographs* No. 46. Washington, D.C.: American Ornithologists' Union, 1991.

James, H. F., and S. L. Olson. "The diversity and biogeography of koa-finches (Drepanidini: *Rhodacanthis*) with descriptions of two new species." *Zoological Journal of the Linnean Society* 144 (2005): 527–541.

James, H. F., and S. L. Olson. "A new species of Hawaiian finch (Drepanidini: *Loxioides*) from Makauwahi Cave, island of Kauai." *The Auk* 123 (2006): 335–344.

James, H. F., T. Stafford, D. Steadman, S. Olson, P. Martin, A. Jull, and P. McCoy. "Radiocarbon dates on bones of extinct birds from Hawaii." *Proceedings of the National Academy of Sciences, USA* 84 (1987): 2350–2354.

Juvik, S. P., and J. O. Juvik, eds. *Atlas of Hawaii*, 3rd ed. Honolulu: University of Hawaii Press, 1998.

Kalakaua, David. *The Legends and Myths of Hawaii, The Fables and Folk-lore of a Strange People.* Rutland, Vt.: Tuttle, 1888 (1972 reprint).

Kikuchi, W. K. "Prehistoric Hawaiian fishponds." *Science* 193 (1976): 295–299.

Kikuchi, W. K., D. L. Kikuchi, and D. Burney. "The archaeological excavations of the Makauwahi Sinkhole Site." Report to Historic Preservation Division, State of Hawaii. Puhi: Kaua'i Community College, 2003.

Kirch, P. V. *Feathered Gods and Fishhooks.* Honolulu: University of Hawaii Press, 1985.

Leopold, A. *A Sand County Almanac, and Sketches Here and There.* New York: Oxford University Press, 1949 (1987 ed.).

Livingstone, D. A. "A lightweight piston sampler for lake deposits." *Ecology* 36 (1955): 137–139.

McGovern, T. H. "Cows, harp seals, and churchbells: Adaptation and extinction in Norse Greenland." *Human Ecology* 8 (1980): 245–273.

MacPhee, R. D. E., ed. *Extinctions in Near Time: Causes, Contexts, and Consequences.* New York: Plenum/Kluwer, 1999.

Malo, D. *Hawaiian Antiquities,* translated by N. Emerson. Honolulu: Bishop Museum Press, 1951 (originally published 1839).

Martin, P. S. "Africa and Pleistocene overkill." *Nature* 212 (1966): 339–342.

Martin, P. S. "Prehistoric overkill: The global model." In *Quaternary Extinctions: A Prehistoric Revolution,* ed. P. S. Martin and R. G. Klein, pp. 354–403. Tucson: University of Arizona Press, 1984.

Martin, P. S. "The last entire earth." *Wild Earth* 2 (1992): 29–31.

Martin, P. S. "Thinking like a canyon: Wild ideas and wild burros." In *Grand Canyon: A Century of Change; Rephotography of the 1889–1890 Stanton Expedition,* by R. H. Webb, pp. 82–83. Tucson: University of Arizona Press, 1996.

Martin, P. S. *Twilight of the Mammoths: Ice Age Extinctions and the Rewilding of America.* Berkeley: University of California Press, 2005.

Martin, P. S. and D. A. Burney. "Bring back the elephants!" *Wild Earth.* Spring 1999: 57–64.

Martin, P. S., and D. W. Steadman. "Prehistoric extinctions on islands and continents." In *Extinctions in Near Time: Causes, Contexts, and Consequences,* ed. R. D. E. MacPhee, pp. 17–55. New York: Plenum/Kluwer, 1999.

Martin, P. S., and H. E. Wright, Jr., eds. *Pleistocene Extinctions: The Search for a Cause.* New Haven: Yale University Press, 1967.

Masse, W. B. and H. D. Tuggle. "The date of Hawaiian colonization." In *Easter Island in Pacific Context: South Seas Symposium; Proceedings of the Fourth International Conference on Easter Island and East Polynesia,* ed. C. M. Stevenson, G. Lee, and F. J. Morin, pp. 229–235. Easter Island Foundation Occasional Paper 4. Los Osos, California: Bearsville and Cloud Mountain Presses, 1998.

Mosimann, J., and P. S. Martin. "Simulating overkill by Paleoindians." *American Scientist* 63 (1975): 304–313.

National Tropical Botanical Garden. *Strategic Directions for the 21st Century: Envisioning the Conservation Potential of the National Tropical Botanical Garden.* Proceedings of Conservation Summit, Kalaheo, Hawaii, April 1–2, 2004.

Olson, S. L., and H. F. James. "Fossil birds from the Hawaiian Islands: Evidence for wholesale extinction by man before Western contact." *Science* 217 (1982): 633–635.

Olson, S. L., and H. F. James. "The role of Polynesians in the extinction of the avifauna of the Hawaiian Islands." In *Quaternary Extinctions: A Prehistoric Revolution,* ed. P. S. Martin and R. G. Klein, pp. 768–780. Tucson: University of Arizona Press, 1984.

Olson, S. L., and H. F. James. *Descriptions of Thirty-two New Species of Birds from the Hawaiian Islands: Part I. Non-Passeriformes.* Ornithological Monographs, no. 45, Washington, D.C.: American Ornithologists' Union, 1991.

Olson, S. L., and H. F. James. "Prehistoric status and distribution of the Hawaiian Hawk (*Buteo solitarius*), with the first fossil record from Kaua'i." *Bishop Museum Occasional Papers* 49 (1997): 65–69.

Palkovacs, E. P., J. Gerlach, and A. Caccone. "The evolutionary origins of Indian Ocean tortoises (*Dipsochelys*)." *Molecular Phylogenetics and Evolution* 24 (2002): 216–227.

Pratt, H. D., P. L. Bruner, and D. G. Berrett. *A Field Guide to the Birds of Hawaii and the Tropical Pacific.* Princeton, N.J.: Princeton University Press, 1987.

Quammen, D. *The Song of the Dodo.* New York: Simon & Schuster, 1996.

Robinson, G. S., L. P. Burney, and D. A. Burney. "Landscape paleoecology and megafaunal extinction in southeastern New York State." *Ecological Monographs* 75 (2005): 295–315.

Selling, O. *Studies in Hawaiian pollen statistics: Part I. The spores of the Hawaiian pteridophytes.* Honolulu: Bishop Museum Special Publication 37, 1946.

Selling, O. *Studies in Hawaiian pollen statistics: Part II. The pollen of the Hawaiian phanerogams.* Honolulu: Bishop Museum Special Publication 38, 1947.

Selling, O. *Studies in Hawaiian pollen statistics: Part III. On the Late Quaternary history of the Hawaiian vegetation.* Honolulu: Bernice P. Bishop Museum Special Publication 39, 1948.

Semonin, P. *American Monster: How the Nation's First Prehistoric Creature Became a Symbol of National Identity.* New York: New York University Press, 2000.

Smith, R. J. "The Pleistocene Park Project—Removing Civilization from North America." Speech from Ninth Annual National Conference of Private Property Rights, Competitive Enterprise Institute, 2005 (www.prfamerica.org/PleistoceneParkProject.html).

Soulé, M. E., ed. *Conservation Biology: The Science of Scarcity and Diversity.* Sunderland, Mass.: Sinauer Associates, 1986.

Stannard, D. E. *Before the Horror: The Population of Hawaii on the Eve of Western Contact.* Honolulu: University of Hawaii Press, 1989.

Steadman, D. W. "Prehistoric extinctions of Pacific island birds: Biodiversity meets zooarchaeology." *Science* 267 (1995): 1123–1131.

Steadman, D. W. *Extinction and Biogeography of Tropical Pacific Birds.* Chicago: University of Chicago Press, 2006.

Steadman, D. W., and P. S. Martin. "The late Quaternary extinction and future resurrection of birds on Pacific islands." *Earth-Science Reviews* 61 (2003): 133–147.

Stolzenburg, W. "Where the wild things were." *Conservation in Practice* 7 (2005): 28–34.

Stone, R. "A bold plan to re-create a long-lost Siberian ecosystem." *Science* 282 (1998): 31–34.

Tanimoto, C. K. *Return to Mahaulepu.* Privately published, 1982.

Tyson, P. *The Eighth Continent: Life, Death, and Discovery in the Lost World of Madagascar.* New York: HarperCollins, 2000.

Vancouver, G., *A Voyage of Discovery to the North Pacific Ocean and Round the World in the "Discovery" and "Chatham," Vols. 1–3 and Atlas of Charts.* London, 1801.

Vitousek, P. M., T. N. Ladefoged, P. V. Kirch, A. S. Hartshorn, M. W. Graves, S. C. Hotchkiss, S. Tuljapurkar, and O. A. Chadwick. "Soils, agriculture, and society in pre-contact Hawaii." *Science* 304 (2004): 1665–1669.

Wagner, W. L., M. Bruegmann, D. R. Herbst, and J. Q. Lau. "Hawaiian vascular plants at risk: 1999." *Bishop Museum Occasional Paper* 60 (1999): 1–64.

Zimov, S. A. "Pleistocene Park: Return of the mammoth's ecosystem." *Science* 308 (2005): 796–798.

Zimov, S. A., V. I. Chuprynin, A. P. Oreshko, F. S. Chapin III, J. F. Reynolds, and M. C. Chapin. "Steppe-tundra transition: A herbivore-driven biome shift at the end of the Pleistocene." *American Naturalist* 146 (1995): 765–794.

Index

Page numbers in italic type indicate illustrations

Adelocosa anops (Kaua'i blind cave wolf spider), 109, *110, 111*
"Africa and Pleistocene Overkill" (Martin), 3
akialoa, 83, Plate 4
Alaka'i Swamp, 69–71
albatrosses, 73, 74
Alekoko Fishpond (Menehune Fishpond), *59,* 59–61
Aleurites moluccana (kukui nut), 44, 66
algae skeletons, 8
Allerton, Robert and John, 57
Allerton Garden, 114, 140
Alyxia stellata (maile vine), 139–140
'ama'ama (*Mugil cephalus,* mullet), 42, 61
Americorps, 98
Andrahomana Cave (Madagascar), 42
Anini Beach, 64, *65,* 66
aragonite, 39
Asio flammeus (Hawaiian short-eared owl), 34–35
Asquith, Adam, 72, 111
Athens, Steve, 63, 69
avian bones, 10–13, 15, 33, 34, 37
avian malaria, 82
avian pox, 82

banyan tree, 13, *14,* 17, 18, Plate 6
basalt mirror, 93, *94*
beads, 94, *94*
Berger, Joel, 161
Bergson, Henri, 104
Bingham, Hiram, 99, *99*
biological invasions, 1, 79, 82–83, 105, 133, 134, 164, 170
birds. *See* fossil bird bones
Bishop Museum Archives, 31, 89, 101, 168
Blackburnia beetle, Plate 4
Blay, Chuck, 39
Bliley, Dan, 24–25, 26, 33
Blitzkrieg Hypothesis, 3, 5, 80
blowholes, 38, 42, 125
boobies, 73, 74
Bock, Carl, 161
Bock, Jane, 161
Boy Scouts, *149,* 149–151
Brunel University conference, 78–79
bucket augers, 15
Burney, Alec, 9, *12, 43,* 64, 90–91, 117
Burney, Lida Pigott, *12,* Plate 2
- and Alaka'i Swamp core, 69–71
- and Carolina Bays excavations, 23–24
- and exploration of Makauwahi sinkhole and cave, 15
- films, 77
- Flat Top Bog analysis, 68–69
- Iliahi House restoration, 150–152
- and lease on Makauwahi Cave, 113–114
- and Limahuli core, 62
- and Makauwahi Cave Reserve, 117, 121
- and Management Unit 2 (Lida's Field of Dreams), 128–129, 144
- and rewilding, 133
- and time travel adventures, 8, 9

Burney, Mara, 9, *12,* 56, 66, 117

Capparis sandwichiana (maiapilo, Hawaiian capers), 108, 121, Plate 2, Plate 4
Carelia spp., 20, 64, *65*, 66, Plate 4
Carpenter, Alan, 74
Case, Steve, 115
catastrophe
  Brunel University conference on, 78–79
  humans as, 75–76, 79–85
  September 11, 2001, attacks, 76–77, 78
  study of, 83–85
  See also extinction—human arrival and
cave pillbug (*Hawaiioscia* cf. *rotundata*), 109
Cellana exarata (ʻopihi), 90–91
charcoal particles, 8, 44, 50, 51, 53, 54, 62, 63, 71, 72
chickens, 30, 36, 44, 48, 82
Chiefess Kamakahelei Middle School, 150
Chinese opium den, 102
Ching, Francis, 60
Chloridops wahi (extinct finch), *35*
Clidemia hirta (Koster's curse), 83
climate change, and extinctions, 4, 5, 7, 39, 68–69
coconut (*Cocos nucifera*), 44
Colocasia esculenta (taro, kalo), 140
conservation
  in situ and *ex situ* strategies, 131–132
  using clues from the past, 130, 134, 137–138
  See also *inter situ* conservation; rewilding
Cook, James, 67, 81
Cookeconcha sp., 66–67
coprophilous fungi (*Sporormiella* spp.), 51, 53–54, 73–74
Cordia subcordata (kou tree), 106–107
Cox, Paul Alan, 49, 114
crabs, 89–90
Crawshaw, Dave, 112
cuesta, 123
Culbertson, Rob, 112
Culex quinquefasciatus (mosquito), 82, 130
Cyanea kuhihewa, 83
Cyclomastra spp., 64, 66

Davis, Owen, 53
Decandido, Bob, 68–69
Deevey, Ed, 167
deforestation, 50, 54, 66, 75
Diamond, Jared, 52
diatoms, 8, 19, 42, 44, 68–69
Dioscorea bulbifera (yam), 91, *93*, 140
DiPietro, Jeri, 112
dodo (*Raphus cucullatus*), 79, 82, 155
dogs, 36, 44, 82
Donlan, Josh, 161, 163, 164
Drosera angliae (sundew), 72
ducks, 54, 75, 80, 131
Duke University, 7–9
dung-fungus spores (*Sporormiella* spp.), 51, 53–54, 73–74
Dye, Tom, 74

"Early Man in Island Environments" (Burney), 10
ecotourism, 141, 157, 165
Endangered Species Act, 131, 168
Endodonta spp., 66
Environmental Leadership Program, 161
eolianite (eolian calcarenite), 39–40
equids, 164
erosion, 30, 81–82
erosion control, 119
Estes, Jim, 161
Eternal Frontier, The (Flannery), 161, 164
Euglandina rosea (rosy wolf snail), 30, 67

Evolution: A Journey into Where We've Been and Where We're Going (WGBH-Boston), 77
"Evolution's Second Chance" (Burney, Steadman, Martin), 160–161
excavation techniques, 15, 23–27, *25*, 33–34, *37*, Plate 1
Exhumation of the Mastodon, The (Peale), *24*
ex situ conservation, 132
extinction
  cascades, 53
  climate change and, 4, 5, 7, 39, 68–69
  hypotheses, 3–7
  on Kaua'i, 130–131
  preventing, 168–171
  vortex, 53
  waves of, 67
—human arrival and
  biological invasions, 82–83
  Europeans, 45–46, 66–67, 81
  on Kaua'i, 1–3, 36–37, 44, 66–67, 75, 83
  landscape transformation, 81–82
  on Madagascar, 9, 50, 53–54
  negative synergy and, 83–85
  overharvesting and overhunting, 79–81
  Polynesians, 80, 81–82

feral livestock, 31–32, 45, 101–102
finch (*Chloridops wahi*), *35*
finches, 126, Plate 4
fire, 8, 78, 94, 95, 96
  and extinctions, 6
  humans and, 6, 44, 51, 53, 74, 81
  natural (prehuman), 44, 71, 72
  See also charcoal particles
fishing gear, 91, *92*
fishponds, 57–62
  Kekupua Fishpond, 61–62
  Lawai Stream, 57–59
  origins of, 56, 60, 61
  types of, *58*
Flannery, Tim, 161, 164
Flat Top bog (Maui), 10, 68–69, 71
flightless birds, 75, 80, 131
Fordham University, 19, 78, 112, 116
Foreman, Dave, 159, 161
fossil bird bones, 10–13, 15, 34, *37*
frigatebird, 74, Plate 4

Gage, Reginald P., II, 20, 64
geese, 54, 75, 80, 139, Plate 4
genealogies, 63, 88
Gillin, Adena, 86, 102
Gillin, Elbert, 29–30, 100
Girl Scouts, *149*, 149–151
goats, 31–32, 101–102
Goetz, Anita, 69
gourd (*Lagenaria siceraria*), 44
Grallistrix auceps (Kaua'i owl), 34, *35*, 75, Plate 4
Grande Caverne (Rodrigues), 157–158
Great Mahele (Land Court Awards), 95
Greene, Harry, 161, 163
Griffiths, Owen, 157–158
groundcovers, establishing, 140
Grove Farm Company
  Iliahi House restoration, 141, *147*, *149*, 149–151
  irrigation, 30–31
  Makauwahi Cave ownership, 18, 34, 95
  Makauwahi Cave Reserve lease, 105, 113–114, 115–116
Guinea grass (*Panicum maximum*), 108

Ha'ena Dry Cave, 73
Hammett, Hal, 61
Ha'upu, 99
Hawaiian capers (*Capparis sandwichiana*, maiapilo), 108, 121, Plate 2, Plate 4

Hawaiian Islands, 1–3, 130–131
Hawaii Conservation Conference (2007), 133
Hawaii Department of Land and Natural Resources, 87
Hawaiioscia cf. *rotundata* (cave pillbug), 109
hawk, Hawaiian (i'o), Plate 4
Heacock, Don, 71
Hearty, Paul, 39
Helping Hands (Na Lima Kokua), 144, 169
hippie commune, 102
Holocene epoch, 68, 69, 71, 137, 142
honeycreepers, 37, 82
horses, 164
Hotchkiss, Sara, 69
Howarth, Frank, 108
Hubbard, Mark, 105
human arrival
  dispersal from Africa, 1, 2
  on Kaua'i, 52, 62–63, 88
  types of clues used in dating, 51
  types of sites needed for dating, 21, 23
  See also extinction—human arrival and
Human Dimensions of Global Change, 20
Hume, Julian, 35, 155, Plate 4
Hurricane 'Iniki, 16–17, 58, 83, 87
Hurricane 'Iwa, 17, 58
hurricanes, 44, 54, 63, 102, 125, 170
Hyde Park mastodon excavation (New York), 25

I, Gabriel, 86, 97
Ice Age climate cycles, 39, 71, 125
Ile aux Aigrettes (Mauritius), 154–158
Iliahi House, 141, *147, 149,* 149–151
'Iniki, 16–17, 83, 87
in situ conservation, 131–132
International Archaeological Research Institute, Inc. (IARII), 69
inter situ conservation, 132–152
  definition and use of term, 132–133
  genetic considerations, 135–137
  goals and purposes of *inter situ* sites, 139–142
  importance of adaptive management, 137–138
  importance of local support, 143–144
  irrigation, 147–148
  islands and, 133
  mix of techniques used, 133–134
  monitoring results, 135, *136*
  opposition and controversy, 142–143, 170
  plant-establishment methods, 145, *146,* 147, *147*
  potential of, 134
  role of paleoecology in, 134, 137–138
  unique challenges of each site, 138–139, 170
  volunteers, 148–152, *149*
  See also rewilding
invasions. *See* biological invasions
i'o (Hawaiian hawk), Plate 4
irrigation channels, 29–30
Island Burial Council, 20
island environments, 9, 10, 80, 133. *See also names of specific islands*

James, Helen, 10–13, *12,* 15, 19, Plate 5
Jefferson, Thomas, 25–26, 166
Jersey Wildlife Trust, 156
Junior Restoration Teams, 141–142

Kahanu Garden, 140
Kalakaua, David, 88
kalo (*Colocasia esculenta,* taro), 140
Kamehameha I, 97
Kanaele (Wahiawa) Bog, 71
Kapaka-Arboleda, LaFrance, 18, 20, 87, 95, 96–98

Kapunakea Pond, 30, 31, 42, 99–101, *100, 101*
Kaua'i
climate change, 68–74
conservation on (see *inter situ* conservation)
extinct and endangered species, 1–3, 130–131, 170
fossil bird bones, 10–13
as Garden Island, 9
human arrival, 1–3, 36–37, 44, 62–63, 75
Hurricane 'Iniki, 16–17, 83, 87
land snail fossil sites, *65*
paleoecological sites, 48, 49–50, 54, *55* (*see also* fishponds; Makauwahi Cave and Reserve)
restoration sites, *55*
See also National Tropical Botanical Garden (NTBG)
Kaua'i blind cave wolf spider (*Adelocosa anops*), 109, *110, 111*
Kaua'i cave amphipod (*Spelaeorchestia koloana*), 108–109, *110, 111*
Kaua'i Community College (KCC), 19, 21, 29, 91, 105, 114, 115
Kaua'i o'o (*Moho braccatus*), *35,* 83, Plate 4
Kaua'i owl (*Grallistrix auceps*), 34, *35,* Plate 4
Keahikuni Kekauonohi, 95, 96, 126
Kealia, 64, 66
Kekupua Fishpond, 61–62
Kenya, Masai Mara region, 3
keystone species, 6, 159
Kikuchi, William K. ("Pila")
community standing of, 18–19
on fishponds, 56, 57, *58,* 62
illness and death of, 86–87
kou trees planted by, 106–107, 143
Māhā'ulepū area research, 96–98, 99–100, 101
and Makauwahi Cave, 13, 21, 29, 31, 87, 91, 95–96
Pila's Point, 118
Killermann, Adam, 121
Kilohana Crater, 71, 141
Kilohana Overlook, 69–70
Kirch, Patrick, 21, 98
Klein, William, 21, 48–49
Kokia kauaensis, Plate 4
Koloa duck (*Anas wyvilliana*), 131
Kostel-Hughes, Faith, 69
Koster's curse (*Clidemia hirta*), 83
kou tree (*Cordia subcordata*), 106–107
Kūkona, Chief, 88–89
kukui nut (*Aleurites moluccana*), 44, 66

Ladderites. *See* Turner conference (2004)
Lagenaria siceraria (gourd), 44
lagerstätten, 125
landscape transformation, 81–82
land snails, 15, 20, 56, 64–67, *65*
"Last Entire Earth, The" (Martin), 159
Late Quaternary Environmental Dynamics of Madagascar (Burney), 9
Lāwa'i-kai, 57, 114, 141, 149
Lawai Stream, 57–59, 141
leadwort (*Plumbago zeylanica*), 123
Lehua Islet, *73,* 74
Leptachatina sp., 66
Lichen Foundation, 161
Lida's Field of Dreams, 128–129, *136,* 139–142, 145–148
Limahuli Gardens, 63, 140, 144
limestone quarry (Māhā'ulepū Quarry), 30, 39, 40, 64, 90, 118, *119*
livestock, feral, 31–32, 45, 101–102
Livingstone, Daniel, 8, 26
Livingstone Sampler, 8
lonomea tree (*Sapindus oahuense*), 126
loulu palm (*Pritchardia* spp.), 111, 118, 121
Loxioides spp. (palila), Plate 4, Plate 5

McCloskey, Dierdre, 44
McNeil, Cameron, 72
MacPhee, Ross, 79, 84
Madagascar, 9, 42, 50, 77, 161
Māhā'ulepū area
drainage and erosion, 30
formation of, 38–39
Hawaiian habitation of, 98–99, *99*
legends, 88–89
meaning of, 96–98, 122
See also Grove Farm Company; Makauwahi Cave and Reserve
Māhā'ulepū Quarry (limestone quarry), 30, 39, 40, 64, 90, 118, *119*
maiapilo (Hawaiian capers, *Capparis sandwichiana*), 108, 121, Plate 2, Plate 4
maile vine (*Alyxia stellata*), 139–140
Makauwahi Cave and Reserve
aerial photo, *119*
banyan tree, 13, *14,* 17, 18, Plate 6
cave formation, 39–41
early exploration and sampling of (1992), *12,* 13–16, *14*
East pit, 47
effects of Hurricane 'Iniki, 16–17
entrance, 31, 42, 100, *101,* 123
European contact, 31–32, 45–46, 126
excavation dangers, 26, 47–48
excavation goals, 21, 23
excavation permissions, 18–20
excavation techniques, 23–27, 33–34, *37,* Plate 1
financial support, 20–21
historical uses of, vii, 91, 93–95, 102–103
history confirmation through other site excavations, 48, 49–50, 54, *55,* 56
human burials in, 19–20
maps, *22, 120*
Milo Patch, 128
modern layer, 29–31, 124–125
name of, 95–96
nineteenth-century sand layer, 31–32
North Cave, 12, *101,* 123–124
ocean incursion and marine layer, 38, 41–42
ownership, 18, 20, 95
painting, Plate 4
Pila's grove of kou trees, 106–107
Polynesian artifacts, 32, 36, 44, 45, 91–95
prehuman plant and animal layer, 36–38
reserve lease and liability insurance, 105–108, 112–117
Reserve Management Unit 2 (Lida's Field of Dreams), 128–129, *136,* 138–141, 145–148
restoration (see *inter situ* conservation)
sediment layers overview, *46,* 124–128
sinkhole, *16, 43,* Plate 6, Plate 7
sinkhole formation, 38, 42, 125
sinkhole overview, 122–123
sinkhole restoration, *106, 107,* 108, 111–112
South Cave, 13, 29, 36, 91, 93, 95, 126–127 (*see also* Troglobite Room)
speleothems, 41, 124, 127
stories and legends, 97–98
as subterranean lake, 42, 44
tours, 123–128, 141
Troglobite Room, 108–111, *110, 111*
tsunami rock deposit, 32–34, 44–45
view from, 118, 121
volcanic layers, 38–39
volunteers, 21, 29, 91, 106, 107, 111–112, 121
weather station and brain center, 121–122, 135
Malama Maha'ulepu, 112, 115–116

Malo, David, 88
mammoths, 53, 160, 166
Martin, Paul, 21
Blitzkrieg Hypothesis, 3, 80
education and career, 167–168
on North American megafauna restoration, 159, 160–161
and rewilding, 166–167
Mascarene Islands, 139, 154–158
mastodons, *24*, *25*, 26, 166
Maui, 10, 68–69, 71
mauka marshes, 68–74
Alaka'i Swamp, 69–71
Flat Top Bog (Haleakala National Park, Maui), 10, 68–69, 71
Kilohana Crater, 71
Silver Falls Bog, 71–72
wetland on Midler's property, 72
Mauritian pink pigeon, 156
Mauritius, 154–158
megafauna, 52–53, 159–161, 162–164
menehune, 60, 61
Menehune Fishpond (Alekoko Fishpond), *59*, 59–61
Midler, Bette, 72
milo (*Thespesia populnea*), 128
moa-nalo, turtle-jawed (*Chelychelynechen quassus*), 75, 80, 139, Plate 4
Moho braccatus (Kaua'i o'o), *35*, 83, Plate 4
mosquito (*Culex quinquefasciatus*), 82, 130
Mount Wai'ale'ale, 40
Mugil cephalus (mullet, 'ama'ama), 42, 61
mullet ('ama'ama, *Mugil cephalus*), 42, 61
Muneno, Kathy, 150

Na Lima Kokua (the Helping Hands), 144, 169
National Museum of Natural History Bird Division, 20–21
National Oceanic and Atmospheric Administration, 20
National Public Radio, 163–164
National Science Foundation, 21, 33, 49
National Tropical Botanical Garden (NTBG), 105
Allerton Garden, 114, 140
Conservation Department, 116–117, 131
Conservation Summit, 114–115
ecotourism, 141
Iliahi House restoration, 151
inter situ projects, 133
Junior Restoration Teams, 141–142
Limahuli Gardens, 63
and Makauwahi Cave, 21, 48–49, 108, 111, 114–115
rare plants, 83, 132, 139
restoration of early Polynesian agricultural varieties, 140
and stopping an extinction event, 168–171
volunteers, 144, 169
Nature, 3, 162–163
Nature Conservancy, 168
negative synergy, 83–85
nene, 131, Plate 4
New Zealand, 42, 154
Niedengarde, Lee, 151
Ni'ihau, 31, 73, 74
9/11 attacks, 76–77, 78
Nishek, Bob, 108, 151
No-eyed Big-eyed Spider. *See* Kaua'i blind cave wolf spider
noho (stool), 94
Noss, Reed, 159

Olson, Storrs
on crabs, 90
Kaua'i research, 12–13, 19

Olson, Storrs (*continued*)
- and Makauwahi Cave excavation, *12*, 13, 15, 29, 30–35, Plate 3, Plate 5
- personality of, 10–11, 24–25
- use of term moa-nalo, 80

Olson, Sydney, 11, *12*
Olson, Travis, 11, *12*
o'o (*Moho braccatus*), *35*, 83, Plate 4
'opihi (*Cellana exarata*), 90–91
O'Rourke, Mary Kay, 167
overharvesting and overhunting, 79–81
overkill hypotheses, 3, 5
owls, 34–35, *35*, 44, 74, 75, Plate 4

Pa'a, 12, 96, 98
Pacific rat (*Rattus exulans*), 36, 44, 51, 62, 82, 138
paleoecology
- and clues to human arrival, 50–54
- importance of site selection, 15, 21, 23
- at landscape level, 38, 49–50
- methods, 7–9
- and restoration, 134, 137–138, 153–154, 159–166

palila (*Loxioides* spp.), Plate 4, Plate 5
palynology, 8, 69
Pancake Rocks (New Zealand), 42
Panicum maximum (Guinea grass), 108
Peale, Charles Willson, *24*, 25, 26
petrel, Hawaiian, Plate 4
petroglyphs, 102
Pezophaps solitaria (solitaire), 158
phallic stone (ule), 97–98
P.H.D. (post-hole digger) method, 33–34
pigs, 36, 44, 70, 79, 82, 83, 94–95, 96
Pila's palila, Plate 5
Pila's Point, 118
Pleistocene epoch, 5–6, 39, 68, 70–71, 159
Pleistocene Overkill, 3
Pleistocene Park (Siberia), 159–160
"Pleistocene Park Project—Removing Civilization from North America" (Smith), 165
Pleistocene rewilding, 159–160, 161–166
Plumbago zeylanica (leadwort), 123
pollen and spores
- Kaua'i fossils, 10, 54, 72, 81 (*see also* Makauwahi Cave and Reserve)
- on Madagascar, 50
- in Makauwahi Cave, 19, 23, 44, 91
- and paleoecology, 51, 53
- restoration and, 135, 140, 141, 142
- Selling and, 69, 71
- techniques, 8, 71, 135

Polynesians, 52, 56, 75, 80, 82–83, 91, *93*, 140
"Poor Man's Time Machine," xi–xii, 9, 21, 34, 123–124, 148, 160, 166
Pratt, Dave, 105
prehistoric, use of term, 88
Pritchardia spp. (fan palm), 44, 111, 118, 121, 126, 138, 170, Plate 4
Property Rights Foundation of America, Inc., 165
property-rights movement, 164–166
Pueo (Hawaiian short-eared owl), 34–35
Puerto Rico, 10

Query, Mark, 108

rabbits, 74, 82
rails, Plate 4
Rainbow Man, 102
Raphus cucullatus (dodo), 79, 155
rat, Pacific (*Rattus exulans*), 36, 44, 51, 82, 138
Ratchener, Ule, 72
Raven, Peter, 115
restoration. See *inter situ* conservation; rewilding
resurrection ecology, 166

Return to Mahaulepu (Tanimoto), 102
rewilding
controversy about, 158–159, 162–166
in Hawaii, 168–171
map of sites of, *155*
in Mauritius and Rodrigues, 154–158
in New Zealand, 154
in North America, 158–159, 161–166
Pleistocene Park (Siberia), 159–160
potential of, 134
and property-rights movement, 164–166
use of term, 133, 159
"Rewilding, Island Style: New Ideas in Inter Situ Conservation" (Burney), 133
Rhyncogonus weevil, Plate 4
Robinson, Aubrey, 31
Robinson, Bruce, 61
Robinson, Guy, 50, 54
Robinson, Keith, 61
Robinson, Warren, 61
Rodrigues, 157–158
Roemer, Gary, 161
rosy wolf snail (*Euglandina rosea*), 30, 67
Russia, Pleistocene Park, 159–160

sandalwood, 95, 101
Sandblom, Marissa, 150
Sapindus oahuense (lonomea tree), 126
Schopenhauer, Arthur, 142–143
Science, 56
sea level changes, 39, 40, 41, 125
Selling, Olaf, 69, 71
September 11, 2001, attacks, 76–77, 78
shearwaters, 73, 74, Plate 4
Sills, Ed, 106, 112
Silver Falls Bog, 71–72
sinkhole, *16*, 38, 42, *43*, 122–123, 125, Plate 6, Plate 7
sinkhole restoration, *106*, *107*, 108, 111–112
Smith, Allan, 105
Smith, Felisa, 161
Smith, Robert J., 165
Smithson, Darlow, 77
Smithsonian Institution, 10, 20
snails, 15, 20, 30, 37, 44, 45, 56, 59, 64–67, *65*, 75, 82, 131
Society for Conservation Biology annual meeting, 160–161
soft release, 133, 156
solitaire (*Pezophaps solitaria*), 158
Soulé, Michael, 53, 133, 135, 159, 161
Spelaeorchestia koloana (Kaua'i cave amphipod), 108–109, *110*, *111*
speleothems, 41, 124, 127
Sporormiella spp. (coprophilous fungi), 53–54
Spouting Horn, 42, 125
Steadman, Dave, 160–161
stone mirror, 93, *94*
subphreatic excavation, 23–27
sundew (*Drosera angliae*), 72
sustainable development, 171

Tangalin, Natalia, 139
Tanimoto, Charles Katsumu, 102
taro (kalo, *Colocasia esculenta*), 140
teal, Laysan, Plate 4
Terry-Bender, Lori, 151
Thespesia populnea (milo), 128
Thoreau, Henry David, 159
time. *See* "Poor Man's Time Machine"
time vertigo, 41
tortoises, 139, 156–158
Troglobite Room, 108–111, *110*, *111*, 127–128
tsunamis and tsunami deposits, 32, 44–45, 54, 63, 72–73, 125
Turner, Reed Beauregard ("Beau"), 161–162

Turner, Ted, 161
Turner conference (2004), 161–166
Turner Endangered Species Fund, 161
turtle-jawed nalo (*Chelychelynechen quassus*), 75, 80, 139, Plate 4
Twilight of the Mammoths (Martin), 166

ule (phallic stone), 97–98
U.S. Fish and Wildlife Service, 114, 131
University of Arizona, Desert Laboratory, 167
University of Hawaii, 168

Vancouver, George, 81
vegetable gardens, 140–141
volcanic eruptions, 38–39

Wahiawa (Kanaele) Bog, 71
Waiopili Heiau, 90
Waipa Farmers' Cooperative, 72–73, 141
Ward, Jerome, 69
Water Conservationist of the Year plaque, 147
wetlands. See *mauka* marshes
WGBH-Boston, 77
Wichman, Charles, Jr. ("Chipper"), 114, 116, 169
Wild Earth, 159, 162, 164
witch coven, 103
wolf rewilding, 162
Wong, Tamara, 139–140
Wood, Ken, 108

yam (*Dioscorea bulbifera*), 91, *93,* 140
Yamamoto, Brian, 150, 151
Yent, Martha, 87
Youth Conservation Corps, 111
Yukimura, JoAnn, 16–17

Zimov, Sergei, 159–160